Stochastic Control Approach to
Futures Trading

Modern Trends in Financial Engineering

ISSN: 2424-8371

This new book series, Modern Trends in Financial Engineering, publishes monographs on important contemporary topics in theory and practice of Financial Engineering. The series' objective is to provide cutting-edge mathematical tools and practical financial insights for both academics and professionals in Financial Engineering. The modern trends are motivated by recent market phenomena, new regulations, as well as new financial products and trading/risk management strategies. The series will serve as a convenient medium for researchers, including professors, graduate students, and practitioners, to track the frontier research and latest advances in the field of Financial Engineering.

Published

Vol. 4 *Stochastic Control Approach to Futures Trading*
by Tim Leung and Yang Zhou

Vol. 3 *Employee Stock Options: Exercise Timing, Hedging, and Valuation*
by Tim Leung

Vol. 2 *Stochastic Drawdowns*
by Hongzhong Zhang

Vol. 1 *Optimal Mean Reversion Trading: Mathematical Analysis and Practical Applications*
by Tim Leung and Xin Li

More information on this series can also be found at http://www.worldscientific.com/series/mtfe

Modern Trends in Financial Engineering | Volume 4

Stochastic Control Approach to
Futures Trading

Tim Leung
Yang Zhou
University of Washington, USA

World Scientific

NEW JERSEY • LONDON • SINGAPORE • BEIJING • SHANGHAI • HONG KONG • TAIPEI • CHENNAI

Published by

World Scientific Publishing Co. Pte. Ltd.
5 Toh Tuck Link, Singapore 596224
USA office: 27 Warren Street, Suite 401-402, Hackensack, NJ 07601
UK office: 57 Shelton Street, Covent Garden, London WC2H 9HE

Library of Congress Control Number: 2024049424

British Library Cataloguing-in-Publication Data
A catalogue record for this book is available from the British Library.

Modern Trends in Financial Engineering — Vol. 4
STOCHASTIC CONTROL APPROACH TO FUTURES TRADING

ISBN 978-981-12-8274-4 (hardcover)
ISBN 978-981-98-0594-5 (ebook for institutions)
ISBN 978-981-98-0704-8 (ebook for individuals)

For any available supplementary material, please visit
https://www.worldscientific.com/worldscibooks/10.1142/13579#t=suppl

Desk Editors: Aanand Jayaraman/Venkatesh Sandhya

Typeset by Stallion Press
Email: enquiries@stallionpress.com

To my parents — Tim Leung

Preface

Futures are standardized exchange-traded bilateral contracts of agreement to buy or sell an asset at a set price on a prespecified future date. Every day, tens of millions of contracts are traded on various futures exchanges around the globe. In recent years, futures have been incorporated into other financial securities, such as commodity and volatility exchange-traded products (ETPs), and have become the driving force behind the price dynamics of these financial instruments.

In the hedge fund industry, managed futures portfolios play an integral role, with hundreds of billions under management. These investments are managed by professionals known as Commodity Trading Advisors (CTAs), and typically involve trading futures on commodities, currencies, interest rates, and other assets. One appeal of managed futures strategies that is often advertised is their potential to produce uncorrelated or superior returns, as well as characteristically different risk-return profiles, compared to the equity market.

As we can see, futures play a major role in the financial market, yet futures trading problems are neither commonly taught in quantitative finance programs nor well researched. This book aims to provide a systematic approach to futures trading and analyze the mathematical problems that arise from trading futures. We consider a number of stochastic markets, including a general multifactor Gaussian framework and regime-switching model. Our multifactor Gaussian framework captures a number of well-known models, like the Schwartz (1997) model and central tendency Ornstein–Uhlenbeck (CTOU) model. Within this framework, we also introduce a new multiscale CTOU model with fast and slow mean reverting diffusion processes. In the regime-switching approach, we analyze the futures trading problem when the underlying asset is driven by the regime-switching geometric Brownian motion or regime-switching exponential Ornstein-Uhlenbeck model. In order to model the stochastic spreads between the underlying spot price and associated futures prices, we introduce a multidimensional scaled Brownian bridge that captures the stochastic convergence in the price behaviors as well as correlation among different underlyings and associated contracts.

We also take a novel stochastic control approach to formulate and solve these trading problems. This common theme leads to the analytical and numerical studies of the associated differential equations in order to derive and compute the optimal trading strategies. In addition, the portfolio optimization problem is incorporated with constraints on the futures position. Our general setup captures the dollar neutral and market neutral constraints, which are widely used in industry.

The trading problem is studied in several different market frameworks. As such, notations and definitions in each chapter are intended for that chapter only. This allows the reader to jump into any chapter of interest and alleviates the reader's burden to look for definitions or notations in earlier chapters.

When writing this book, we aim to make it useful not only for graduate students and researchers, but also practitioners who specialize in trading futures and related derivatives. In particular, numerical implementation is discussed and illustrative examples are provided.

We would like to express our gratitude to a number of people who have helped make this book project possible. This book was motivated by a series of projects on futures pricing and trading previously completed with my collaborators, including Raphael Yan, Bahman Angoshtari, Xiaodong Chen, Xin Li, Zheng Wang, and Brian Ward. Graduate students in the Computational Finance and Risk Management (CFRM) Program at University of Washington (UW) have participated in our lectures on dynamic futures portfolio optimization, and some have taken part in the early reading of the drafts of various chapters. The helpful remarks and suggestions by the editorial board of this book series are also greatly appreciated. We also thank the participants at the AMS Meeting, American Control Conferences, and SIAM Conference on Financial Mathematics & Engineering for the comments on this work.

Lastly, we want to give our special thanks to Ms. Rochelle Kronzek of World Scientific. Without her encouragement and patience over the years, we would never have started nor completed this book project.

Tim Leung
Boeing Endowed Chair Professor
Department of Applied Mathematics
University of Washington
Seattle, WA

July 2024

About the Authors

Tim Leung is the Boeing Endowed Chair Professor of Applied Mathematics and the Director of the Computational Finance and Risk Management (CFRM) program at the University of Washington in Seattle. Previously, he had been a professor at Johns Hopkins University and Columbia University. He obtained his Ph.D. in Financial Engineering and Operations Research from Princeton University. He has published over 70 peer-reviewed articles, along with several books on the topics of mean reversion trading, ETFs, futures trading, and employee stock options. Professor Leung serves on the advisory board for the AI for Finance Institute as well as the editorial board of multiple journals on quantitative finance.

Yang Zhou received his Ph.D. in Applied Mathematics from the University of Washington. His research areas include financial mathematics, futures trading, and stochastic control. He has published several journal articles on dynamic futures portfolios, employee stock options valuation, and regime-switching models. At present, he is a Quantitative Analytics Specialist at Wells Fargo, focusing on counterparty risk and advanced risk models.

Contents

List of Figures

List of Tables

Chapter 1

Introduction

This chapter gives an overview of the futures market, followed by a discussion on futures price structure. This allows us to better under the price dynamics of futures and motivate a stochastic control approach to futures trading. An introduction to the subsequent chapters is also provided.

1.1 Futures Market

Futures are standardized exchange-traded bilateral contracts of agreement to buy or sell an asset at a predetermined price at a prespecified time in the future. At the Chicago Mercantile Exchange (CME), futures contracts are offered for a wide array of underlying assets. A sample list of futures categories and underlying assets is provided in Table 1.1. Different futures markets have different and varying trading volumes. By recent market observations, the following are the most heavily traded futures contracts: Eurodollar (GE), E-Mini S&P 500 (ES), Crude Oil WTI (CL), 10-Year Treasury Note (ZN). Over the past several years, the trading volume on CME has averaged well over ten million contracts per day[1] and new contracts, such as cryptocurrency futures, have been introduced to the market.

The futures market allows commodities producers and consumers to hedge against adverse commodity price movements. These important market participants are commonly called the hedgers. They include farmers, wholesalers, manufacturers, and retailers. Utilizing futures contacts, they can potentially limit the risk of financial losses.

[1] *Source*: The CME Group daily exchange volume and open interest report.

Table 1.1. Futures markets and examples of underlyings for contracts traded at the CME as of 2024.

Interest rates	Foreign exchange	Indices
2-Year T-Note	Euro	S&P 500
5-Year T-Note	British pound	Nasdaq 100
10-Year T-Note	Japanese yen	Dow Jones
US Treasury Bond	Australian dollar	Russell 2000
1-Month SOFR	Canadian dollar	S&P 500 VIX
3-Month SOFR	Swiss franc	S&P GSCI
30-Day Fed Funds	New Zealand dollar	Nikkei
1-Month SONIA	Swedish krona	
3-Month SONIA	Norwegian krone	
	Mexican peso	
	Brazilian real	
Energy	**Metals**	**Grains**
WTI crude oil	Gold	Corn
Brent crude oil	Silver	Wheat
Natural gas	Copper	Soybean
Gasoline	Aluminum	Soybean oil
Heating oil	Platinum	Soybean meal
	Palladium	Oats
Livestock & dairy	**Softs**	**Crpytocurrencies**
Live cattle	Sugar	Bitcoin
Feeder cattle	Cocoa	Ether
Lean hogs	Coffee	
Milk	Cotton	
Butter	Orange juice	
Cheese	Lumber	

For example, a wheat farmer who is worried about lower future wheat prices can mitigate price risk by selling wheat futures contracts. This guarantees that the farmer will receive a predetermined price for the wheat delivery. Hedgers who want to guard against falling commodity prices are called the sell-side hedgers. On the other hand, a cereal producer who is a large consumer of wheat may want to lock into a fixed price of wheat now by buying wheat futures in view of market uncertainty. Concerned about rising commodity prices, the cereal producer is a buy-side hedger. In addition, there are also merchandisers who may buy and sell commodities. This may result in different directional risks as compared to pure buy-side or sell-side hedgers.

While hedgers use futures contracts to mitigate the price risk in commodities, speculators seek to trade these contracts to profit from favorable price movements. In addition, there are arbitrageurs who will buy or sell futures and the underlying commodities when profitable arbitrage opportunities arise. In modern futures markets, there are numerous funds with direct investment in commodity futures and options, as well as over-the-counter (OTC) swap dealers hedging commodity index exposure.

Most futures contracts are not held until expiry. In fact, they are typically liquidated through offsetting contracts by traders. In such cases, there is no delivery of commodities. Although many contracts do not result in delivery, the delivery provision in each futures contract ensures convergence between the futures and spot prices.

In order to take a futures position, one does not need to have the full cash amount of the contract value. Instead, a margin deposit, which is only a small portion of the contract value, is required. This renders futures a leveraged instrument, allowing investors to potentially realize significant returns and losses.

1.2 Futures Price Dynamics

For the same underlying asset, there are multiple futures contracts with different expiration dates. The spot price is the current price of the underlying asset. A common way to interpret futures prices is that they reflect the market's expectation of the underlying asset price in the future.

When the market expects the underlying asset to be more expensive in the future than the current price, then the futures market is said to be in contango. A contango market means that the futures contracts are trading at a premium to the spot price. Since the futures price should converge to the future spot price toward expiry, contango suggests that the futures prices are expected to decrease over time. In contrast, backwardation is when the future price of the underlying asset is anticipated to be less expensive than the spot price. In this scenario, the market may expect the futures prices to increase towards the spot price over time as expiration nears.

In the agricultural commodity markets, a major factor for price variations in many futures markets is seasonality. We can see this in Figure 1.1 taken from Guo and Leung (2017). They analyze the behavior of the basis, which is the difference between the futures price and spot price, in the agricultural futures markets. As shown for all three grains (corn, soybeans,

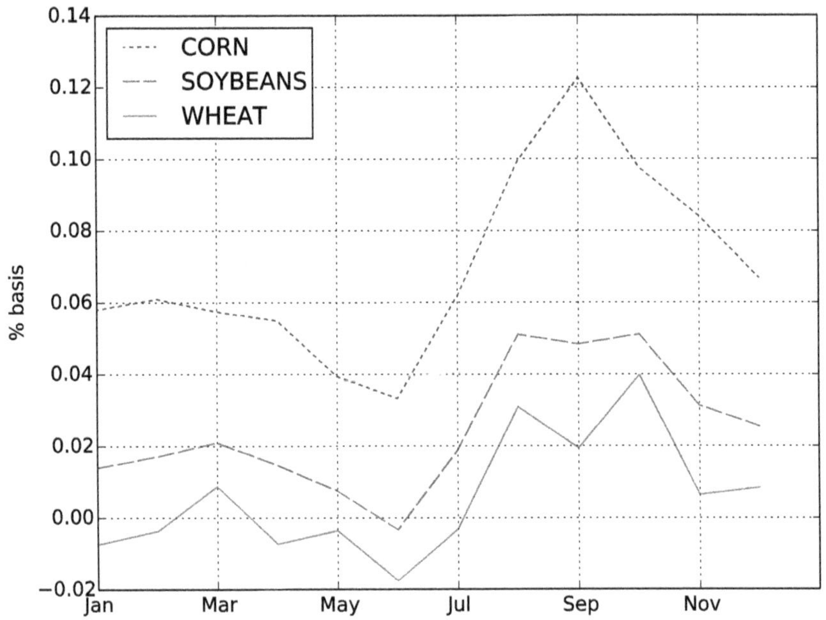

Fig. 1.1. Seasonality of the average basis for corn, soybeans, and wheat. The basis is highest during the harvest months (August–October) as storage rates are high due to grain silo capacity constraint. For instance, the average basis for soybeans exceeds 12% in September. From February to June, the basis tends to be smaller since the storage costs are lower due to empty grain silos before the next harvest begins.

Source: Guo and Leung (2017).

and wheat), the basis is typically higher during the fall harvest months (August–October) when available storage capacity is low and market storage rates are high. In fact, the value of storage is typically higher in these months. In contrast, the basis is typically smaller from the winter through the summer while the grain silos are being emptied before the next harvest, so storage rates are reduced during the low season.

1.3 Managed Futures Portfolios

Managed futures portfolios play an integral role in hedge fund and alternative investment industries. With hundreds of billions under management, these portfolios are managed by professional investment individuals or management companies traditionally known as Commodity Trading Advisors (CTAs). Regulated and monitored by both government agencies such as the US Commodity Futures Trading Commission (CFTC) and National

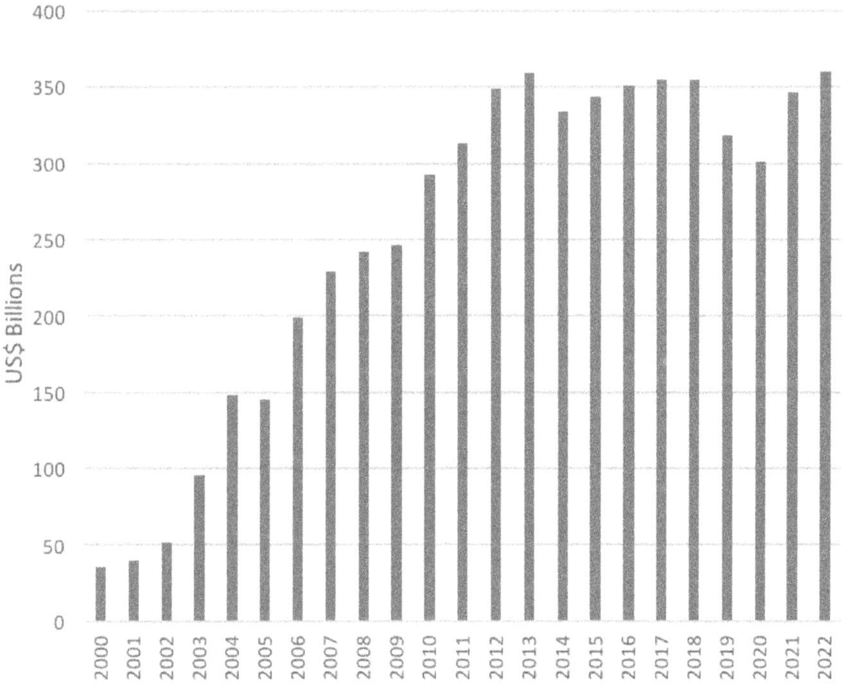

Fig. 1.2. Commodity Trading Advisors (CTAs) asset under management (AUM) from 2000 to 2022.

Source: Barclayhedge Managed Futures AUM data.

Futures Association (NFA), this class of assets has seen significant growth in the past two decades and averaged well over US$300 billion annually during 2011–2022 as shown in Figure 1.2. In addition, futures may also be the core instruments in many exchange-traded funds (ETFs) and exchange-traded notes (ETNs). Examples include the United States Oil Fund (ticket: USO) and Invesco DB Oil fund (ticket: DBO), which hold oil futures continuously in an attempt to track a benchmark. These exchange-traded products may be included in the portfolios of individual and institutional investors.

Among fund managers and traders, the approaches to building their proprietary trading systems are highly varied. Some may build systematic or computer-driven strategies, while others may rely on discretionary methods. For example, a trend-following CTA may seek to capture longer-term trends in the markets, which may involve holding positions from several weeks to over a year. A contrasting approach may speculate on imminent price reversal, which typically involves shorter-term positions and more frequent trades.

The data and signals used in these strategies may also vary greatly. Some managers may primarily use price data and derive signal using machine learning or statistical analysis whereas others may more heavily rely on macroeconomic data and other market fundamentals.

Managed futures programs generally take long or short positions in futures contracts on commodities, currencies, interest rates, and other assets, that are available on futures exchanges (see Table 1.1). With sufficient liquidity and price transparency, the value of a futures portfolio can be measured dynamically as market prices fluctuate. This practice of *mark to market* reflects the portfolio value based on prevailing market prices.

One appeal of managed futures strategies is their advertised potential to produce less correlated or superior returns, as well as different risk-return profiles compared to equities (see Bodie and Rosansky (1980); Gregoriou *et al.* (2010); Elaut *et al.* (2016); Miffre (2016); Demiralay *et al.* (2019); Zaremba *et al.* (2021); Wang *et al.* (2022), among others). For instance, during inflationary periods when returns from equities and bonds are lagging, commodities futures tend to gain popularity among fund managers. Some also seek to capture the unique price dynamics, such as volatility spillover, mean reversion, momentum, as well as arbitrage opportunities, within specific markets (see Richie *et al.* (2008); Cummings and Frino (2011); Trujillo-Barrera *et al.* (2012); Lubnau and Todorova (2015); Smales (2016); Tsuji (2018); Baur and Smales (2022)).

In many managed futures portfolios, futures from different asset classes of different maturities and leverage are often used. The classes of strategies are conceivably diverse among managed futures funds, with the popular ones being long-short strategy and momentum strategy (see Kurov (2008); Hurst *et al.* (2013), among others). It is also possible to combine managed futures with portfolios of other asset classes, which can be viewed as a way to enhance the risk-adjusted returns.

1.4 Stochastic Control Approach to Futures Trading

The primary objective of this book is to present a stochastic control approach to dynamic futures trading. This involves the studies of a number of dynamic futures portfolio optimization problems under several different stochastic market frameworks. The proposed stochastic models are suitable for a wide array of futures markets. The optimal strategies are analytically derived and numerically implemented with illustrations. Trading strategies and portfolio characteristics under different market conditions are also examined.

Before considering the portfolio optimization problem, one crucial step is to establish a stochastic model to adequately represent the price dynamics within the futures market. There are two major contrasting approaches. One way is to introduce a stochastic model for the underlying asset price process. Then, derive the futures prices using the no-arbitrage principle. For example, the futures prices can be computed by taking expectations under some risk-neutral pricing measure. This approach specifies a structural arbitrage-free relationships between the spot and futures prices.

On the other hand, it is possible to introduce a multidimensional stochastic model to represent the joint dynamics of the spot price and multiple stochastic basis processes corresponding to futures contracts with different expiration dates. Then, the futures price is derived from the simple arithmetic relationship between the spot price and basis. This approach is sometimes called the stochastic basis approach (see Angoshtari and Leung (2019a); Chen *et al.* (2022), and references therein).

In contrast to the first approach, the no-arbitrage principle is not employed to generate futures prices, though it does not imply there are arbitrage opportunities. Moreover, the stochastic basis approach, depending on the model specifications, allows for a wider range of possible price behaviors, such as non-comoving futures prices and non-convergence at expiry, as well as other correlation structures, that are commonly observed in some markets.

In Chapter 2, we propose a general multifactor Gaussian framework. Some existing simpler models like the two-factor Schwartz model and central tendency Ornstein–Uhlenbeck (CTOU) model have several desirable features such as their tractability and interpretability. However, they are often inadequate in fitting observed prices of all traded contracts, as pointed out by Carmona and Ludkovski (2004), among others. The multifactor Gaussian framework is a generalization of the two-factor models. It provides additional flexibility and permits good fit to entire term structure as displayed in the market. Especially in deep and liquid futures markets, such as crude oil or gold, with over ten contracts of various maturities actively traded at any given time, multifactor models are particular useful. These models also yield explicit formulas for the futures prices, so they are highly tractable and easy to implement. Through a series of numerical examples, we illustrate the price behaviors of the futures contract under this framework.

In the literature, Cortazar and Naranjo (2006) and Cortazar *et al.* (2008, 2019) apply a multifactor Gaussian model for pricing commodity futures. Cortazar *et al.* (2017) introduce a multifactor stochastic volatility model to enhance calibration against observed commodity option prices. For more general multifactor models for commodity derivatives pricing, we refer to Gómez-Valle and Martínez-Rodríguez (2013) and references therein.

Our proposed Gaussian framework can capture both mean-reverting and trending price dynamics. Moreover, our framework subsumes a number of well-known models, like the Schwartz (1997) model, CTOU model, and several other two-factor models (e.g. Schwartz and Smith, 2000). It also leads us to introduce a new multiscale CTOU model that is driven by a fast and slow mean-reverting process. In particular, this allows us to derive futures prices when the underlying asset has a stochastic long-run mean that is driven by a fast-moving and/or slow-moving factor.

There are empirical studies on testing whether commodity prices following a mean-reverting process (see, e.g. Meade (2010)). As studied by Leung *et al.* (2020), it is also possible to identify the most mean-reverting pairs from a large collection of assets. Lee *et al.* (2023) propose a framework for constructing diversified portfolios with multiple mean-reverting trading strategies.

In Chapter 3, we discuss the optimal dynamic futures trading strategies under the Gaussian framework discussed in Chapter 2. The main objective is to construct and optimize a dynamic portfolio of futures when the underlying asset is driven by a number of Gaussian factors. The optimal futures trading strategy is determined by solving a stochastic control problem, whose objective is to maximize the expected utility from trading wealth. By analyzing and solving the associated Hamilton–Jacobi–Bellman (HJB) equations, we present the value function and optimal trading strategies explicitly.

Asset prices are often considered to be dependent on market conditions. Market regimes may change suddenly and persist for a period of time. The unpredictability of the timing of regime changes also means that the associated risks are almost impossible to hedge. This has critical implications to derivatives pricing and trading. In order to capture these crucial properties of market dynamics, one major approach is to represent stochastic market regimes by a finite-state Markov chain. In most cases, the Markov chain is an exogenous continuous-time random process and is not directly traded. The effects of the Markov chain are reflected in the asset price dynamics. In particular, the asset's expected return and volatility may vary across regimes. In turn, the prices of futures and other derivatives will be affected by the market regimes as well.

Regime switching models date back to Hamilton (1989), who introduced a regime-switching time series model to capture the movements of the stock prices and showed that the regime-switching model represents the stock returns better than the model with deterministic coefficients. Thereafter, a variety of regime-switching models have been applied to many problems in economics and finance, including derivatives pricing (see, e.g. Buffington and Elliott (2002), Elliott *et al.* (2005), and Almansour (2016), among many others), and portfolio selection (see Zhou and Yin (2003); Sass and Haussmann (2004); Leung (2010); Çanakoğlu and Özekici (2012); Capponi and Figueroa-López (2012); Chen *et al.* (2019); Sotomayor and Cadenillas (2009)).

In Chapter 4, we present a general regime-switching framework in which the stochastic market regime is represented by a continuous-time finite-state Markov chain. The underlying asset's spot price is modeled by a Markov-modulated diffusion process. Then, we derive the no-arbitrage price dynamics for the futures contracts. This general framework captures a number of regime-switching models, including the regime-switching geometric Brownian motion (RS-GBM) and regime-switching exponential Ornstein–Uhlenbeck (RS-XOU) model. We study the futures pricing problem and present the futures pricing dynamics.

In Chapter 5, we study the problem of dynamically trading futures in a general regime switching market. After obtaining the no-arbitrage price dynamics for the futures contracts, we determine the optimal futures trading strategy by solving a utility maximization problem. By analyzing and numerically solving the associated HJB equations, we present the value function and optimal trading strategies explicitly for investors. Besides, we define the investor's certainty equivalent to quantify the value of the futures trading opportunity to the investor. In particular, we develop a new three-factor model, the multiscale CTOU process, which has two stochastic mean factors, one fast and one slow. We provide the numerical examples to examine model parameters for our new model.

In comparison to related studies on portfolio optimization with regime switching, we investigate the dynamic trading of futures, rather than stocks, in a general regime-switching market framework that can be applied to an array of regime-switching models. This different setup leads to the interaction between the historical measure and risk-neutral pricing measure, and how they affect the evolution of futures prices and portfolio wealth.

In particular, the regime-switching feature also leads to jumps in the futures prices and the investor's wealth process, even though the underlying asset price has continuous paths. Moreover, under our regime-switching framework, the certainty equivalent is the same across different underlying models as long as the market price of risk stays the same.

In the models discussed above, the futures portfolios consist of multiple futures contracts written on the same underlying asset. A natural extension is to include futures of different underlying assets. This will drastically expand the set of trading instruments and increase the diversity of exposure. As a result, the portfolio manager has access to various assets within the same asset class (e.g. gold, silver, and copper) as well as across asset classes (e.g. precious metals, agricultural commodities, oil and gas, equity, and volatility).

However, such a portfolio expansion comes with some new challenges. With additional underlying assets and futures, the dimension of the portfolio optimization problem is significantly increased. Furthermore, one must address the dependency structure among the underlying assets and their futures, even if the underlying assets are not traded. This calls for a stochastic model that can capture the correlation among all the futures and underlying assets while maintaining analytical tractability, numerically efficiency, as well as interpretability.

Motivated by these observations, we introduce in Chapter 6 an alternative way to directly model the joint price dynamics of the underlying assets and associated futures. Specifically, the stochastic spreads between the underlying spot price and associated futures prices are represented by a multidimensional scaled Brownian bridge.

In practice, market frictions and inefficiencies may render the futures price different from the spot price prior to maturity. For each futures contract, the spread between the two prices is called the *basis*. By no-arbitrage theory, futures prices must converge to the spot price at expiry, so the basis process is expected to converge to zero as the associated futures contract expires as well.

This behavior of the basis process leads us to (i) express each futures price process through the associated basis, and (ii) model the random basis using a Brownian bridge. With multiple futures contracts, we apply a *multidimensional* Brownian bridge, where each component converges to zero at the respective maturity. In addition, the portfolio optimization problem is incorporated with constraints on the futures position. Our general setup captures the dollar neutral and market neutral constraints, which are widely used in industry.

In the literature, evidence of futures basis has motivated several studies on related trading strategies. Simon and Campasano (2014) examine the volatility index (VIX) futures basis and discuss trading strategies involving VIX futures and S&P index futures. Buetow and Henderson (2016) present a link between the option market frictions to changes in the VIX futures basis and find that market information embedded in illiquid S&P options is reflected in the VIX but not in VIX futures.

Several studies have modeled the stochastic basis for trading purposes. For example, Brennan and Schwartz (1988, 1990) assume that the basis of an index futures follows a Brownian bridge. Using an optimal stopping approach, they compute the value of the timing option to trade the basis, along with the optimal trading boundaries. Also under a standard Brownian bridge model, Dai *et al.* (2011) provide an alternative trading strategy and specification of transaction costs. Another related work by Liu and Longstaff (2004) assume that the basis follows a scaled Brownian bridge and the investor is subject to a collateral constraint, and derive the strategy that maximizes the expected logarithmic utility of terminal wealth.

Our model presented in Chapter 6 is based on the authors' companion paper Chen *et al.* (2022). It is related to a number of studies in finance involving Brownian bridges. Applications include modeling the flow of information in the market. For example, Brody *et al.* (2008) use a Brownian bridge as the noise in the information about a future market event. They derive several option pricing formulas based on this asset price dynamics and market information flow. For algorithmic trading, Cartea *et al.* (2016) utilize a randomized Brownian bridge (rBb) to model the mid-price of an asset with a random end-point perceived by an informed investor, and determined the optimal placements of market and limit orders. Leung *et al.* (2018) also apply a similar rBb framework to derive the optimal timing strategies for options trading.

Our study on futures trading has also motivated a number of potential research directions, ranging from new market models for futures pricing to alternative optimization methods for dynamic portfolios.

In various aspects of the futures portfolio optimization problem, a variety of machine learning methods can potentially be useful. For example, Guijarro-Ordonez *et al.* (2022) develop a deep learning framework for statistical arbitrage. For futures trading, Waldow *et al.* (2021) examine a learning-based statistical arbitrage strategy and train several machine learning models to predict price movements.

Increasingly futures are used to construct new exchanged-traded products, such as commodity exchange-traded notes (ETNs). It is most commonly designed to track a commodity index. There are also leveraged versions of these products. Their theoretical long-term performance has been analyzed in Leung and Park (2017) and Leung *et al.* (2023) under a wide array of market models. One can also try to replicate the index using a dynamic or static portfolio of futures. For more details on this topic, we refer to Leung and Ward (2015) and Leung and Ward (2022).

Chapter 2

Futures Pricing Under a Multifactor Gaussian Framework

In this chapter, we discuss no-arbitrage pricing of futures contracts in a general stochastic framework. We propose a multifactor framework where the underlying asset price is driven by a multiple Gaussian diffusion processes. Our framework is flexible and practically useful for pricing futures contracts. The flexibility of multifactor models permits good fit to entire term structure as displayed in the market. These models also come with explicit formulas for the futures prices, so they are highly tractable and easy to implement. It can be viewed as a generalization of some existing factor models, such as the Schwartz (1997) two-factor model and central tendency Ornstein–Uhlenbeck (CTOU) model (see Mencia and Sentana (2013)).

We describe the general market framework and futures dynamics in Section 2.1. The application to the two-factor models, along with numerical implementation and examples, is discussed in Section 2.2. In Section 2.3, we introduce a new multiscale CTOU model that is driven by a fast and slow mean-reverting process. In this model, the price of the underlying asset has a stochastic mean that is the sum of two mean-reverting processes with a fast and slow rate of mean reversion respectively. Numerical examples are presented in Section 2.4 to illustrate the futures prices under this new multiscale model.

The futures pricing formulas for the multifactor Gaussian framework derived herein will be applied in the futures portfolio optimization in Chapter 3. Their explicit form will be crucial in making the portfolio optimization very tractable and easy to implement.

2.1 Multifactor Gaussian Model

We consider a multifactor market model. Let $\boldsymbol{X}_t$ be an N-dimensional vector $(X_t^{(1)}, \ldots, X_t^{(N)})^\top$, where $X^{(1)}$ is the log-price of the underlying asset and $(X^{(2)}, \ldots, X^{(N)})$ are observable stochastic factors. The spot price of the underlying asset is defined by

$$S_t = \exp\left(X_t^{(1)}\right). \tag{2.1}$$

Under physical probability measure $\mathbb{P}$, the vector of random factors $\boldsymbol{X}_t$ evolves according to the vector-valued linear stochastic differential equation (SDE):

$$d\boldsymbol{X}_t = (\boldsymbol{\mu} - \boldsymbol{K}\boldsymbol{X}_t)dt + \boldsymbol{\Sigma} d\tilde{\boldsymbol{Z}}_t^{\mathbb{P}}, \tag{2.2}$$

for $t \geq 0$ with $\boldsymbol{X}_0 \in \mathbb{R}^N$.

Let us discuss the coefficients and parameters in (2.2). In the drift term, $\boldsymbol{\mu} = (\mu_1, \ldots, \mu_N)^\top$ is an N-dimensional column vector of constants, and $\boldsymbol{K}$ is an $N \times N$ matrix with constant entries k_{ij}:

$$\boldsymbol{K} = \begin{bmatrix} k_{11} & k_{12} & \ldots & k_{1N} \\ k_{21} & k_{22} & \ldots & k_{2N} \\ \vdots & \vdots & \ddots & \vdots \\ k_{N1} & k_{N2} & \ldots & k_{NN} \end{bmatrix}.$$

In the diffusion term of (2.2), $\boldsymbol{\Sigma}$ is an $N \times N$ diagonal matrix with constant volatility parameters σ_i for $i = 1, \ldots, N$. Precisely, we have

$$\boldsymbol{\Sigma} = \begin{bmatrix} \sigma_1 & 0 & \ldots & 0 \\ 0 & \sigma_2 & \ldots & 0 \\ \vdots & \vdots & \ddots & \vdots \\ 0 & 0 & \ldots & \sigma_N \end{bmatrix}.$$

The source of randomness that drives SDE (2.2) is an N-dimensional column vector of Brownian motions, denoted by $\tilde{\boldsymbol{Z}}_t^{\mathbb{P}} = (\tilde{Z}_t^{\mathbb{P},1}, \ldots, \tilde{Z}_t^{\mathbb{P},N})^\top$. The Brownian motions are correlated such that vector Brownian motion, $\tilde{\boldsymbol{Z}}_t^{\mathbb{P}}$, satisfies

$$(d\tilde{\boldsymbol{Z}}_t^{\mathbb{P}})(d\tilde{\boldsymbol{Z}}_t^{\mathbb{P}})^\top = \boldsymbol{\Omega}\, dt.$$

Here, $\boldsymbol{\Omega}$ is a matrix whose (i, j)-element of the is the instantaneous correlation $\rho_{ij} \in (-1, 1)$.

We assume that any Brownian motion in $\tilde{\boldsymbol{Z}}_t^{\mathbb{P}}$ could not be replicated by other $N-1$ Brownian motions, which indicates that $\boldsymbol{\Omega}$ is a symmetric positive definite matrix. Therefore, by applying Cholesky decomposition to $\boldsymbol{\Omega}$, we can write

$$\boldsymbol{\Omega} = \boldsymbol{C}\boldsymbol{C}^\top,$$

where $\boldsymbol{C}$ is an invertible lower triangular matrix:

$$\boldsymbol{C} = \begin{bmatrix} c_{11} & 0 & \dots & 0 \\ c_{21} & c_{22} & \dots & 0 \\ \vdots & \vdots & \ddots & \vdots \\ c_{N1} & c_{N2} & \dots & c_{NN} \end{bmatrix}.$$

Thus, we can define

$$d\boldsymbol{Z}_t^{\mathbb{P}} = \boldsymbol{C}^{-1} d\tilde{\boldsymbol{Z}}_t^{\mathbb{P}},$$

which is an N-dimensional column vector of independent Brownian motion increments under measure $\mathbb{P}$. Accordingly, the SDE for $\boldsymbol{X}_t$ can be written as

$$d\boldsymbol{X}_t = (\boldsymbol{\mu} - \boldsymbol{K}\boldsymbol{X}_t)dt + \boldsymbol{\Sigma}\boldsymbol{C}\, d\boldsymbol{Z}_t^{\mathbb{P}}. \tag{2.3}$$

In order to understand the pricing of futures contracts, we need to introduce the risk-neutral pricing measure. Under this measure, futures prices are computed. Let us denote the risk-neutral pricing measure by $\mathbb{Q}$.

The two probability measures $\mathbb{Q}$ and $\mathbb{P}$ are different but connected. Indeed, they are linked via the risk premium vector, denoted by

$$\boldsymbol{\zeta} = (\zeta_1, \dots, \zeta_N)^\top.$$

Precisely, the $\mathbb{Q}$-Brownian motion is related to the $\mathbb{P}$-Brownian motion through the equation

$$d\boldsymbol{Z}_t^{\mathbb{Q}} = d\boldsymbol{Z}_t^{\mathbb{P}} + \boldsymbol{\zeta} dt, \tag{2.4}$$

where $\boldsymbol{Z}_t^{\mathbb{Q}} = (Z_t^{\mathbb{Q},1}, \dots, Z_t^{\mathbb{Q},N})^\top$ is the N-dimensional column vector of independent $\mathbb{Q}$-Brownian motions.

Remark 2.1. The risk premium can potentially be time-varying or stochastic (see e.g. Bhar and Lee (2011)). We have chosen a constant risk premium vector for simplicity and to preserve the same stochastic structure under the physical and risk-neutral probability measures.

Then, under the pricing measure $\mathbb{Q}$, the vector of random factors $\boldsymbol{X}_t$ evolves according to

$$d\boldsymbol{X}_t = (\boldsymbol{\mu} - \boldsymbol{\lambda} - \boldsymbol{K}\boldsymbol{X}_t)dt + \boldsymbol{\Sigma}\boldsymbol{C}\, d\boldsymbol{Z}_t^{\mathbb{Q}}, \tag{2.5}$$

where the N-dimensional column vector $\boldsymbol{\lambda} = (\lambda_1, \ldots, \lambda_N)^\top$ depends on the risk premium vector $\boldsymbol{\zeta}$ and is given by

$$\boldsymbol{\lambda} = \boldsymbol{\Sigma}\boldsymbol{C}\boldsymbol{\zeta}.$$

By observing the SDE (2.5), the volatility and correlation parameters can now affect the drift term of $\boldsymbol{X}_t$ via the vector $\boldsymbol{\lambda}$.

Now we consider a futures contract of maturity T written on the underlying asset S. With the asset price defined in (2.1), the futures price at time $t \in [0, T]$ is given by the risk-neutral expectation of the future price of the underlying asset. Precisely, we write

$$F(t, \boldsymbol{x}) := \mathbb{E}^{\mathbb{Q}}\big[\exp(X_T^{(1)}) \,|\, \boldsymbol{X}_t = \boldsymbol{x}\big]. \tag{2.6}$$

Our next step is to derive the partial differential equation (PDE) associated with the futures price function in (2.6). To that end, we first define the linear differential operator

$$\mathcal{L}^{\mathbb{Q}} \cdot = (\boldsymbol{\mu} - \boldsymbol{\lambda} - \boldsymbol{K}\boldsymbol{x})^\top \nabla_{\boldsymbol{x}} \cdot + \frac{1}{2}\operatorname{Tr}(\boldsymbol{\Sigma}\boldsymbol{\Omega}\boldsymbol{\Sigma}\nabla_{\boldsymbol{x}\boldsymbol{x}} \cdot),$$

where

$$\nabla_{\boldsymbol{x}} \cdot = (\partial_{x_1} \cdot, \ldots, \partial_{x_N} \cdot)^\top$$

is the nabla operator. Also, the Hessian operator $\nabla_{\boldsymbol{x}\boldsymbol{x}} \cdot$ satisfies

$$\nabla_{\boldsymbol{x}\boldsymbol{x}} \cdot = \begin{bmatrix} \partial^2_{x_1} \cdot & \partial_{x_1 x_2} \cdot & \cdots & \partial_{x_1 x_N} \cdot \\ \partial_{x_1 x_2} \cdot & \partial^2_{x_2} \cdot & \cdots & \partial_{x_2 x_N} \cdot \\ \vdots & \vdots & \ddots & \vdots \\ \partial_{x_1 x_N} \cdot & \partial_{x_2 x_N} \cdot & \cdots & \partial^2_{x_N} \cdot \end{bmatrix}.$$

Then, the futures price function $F(t, \boldsymbol{x})$ solves the following PDE:

$$(\partial_t + \mathcal{L}^{\mathbb{Q}})F(t, \boldsymbol{x}) = 0, \tag{2.7}$$

for $(t, \boldsymbol{x}) \in [0, T) \times \mathbb{R}^N$, with the terminal condition

$$F(T, \boldsymbol{x}) = \exp(\boldsymbol{e}_1^\top \boldsymbol{x})$$

for $\boldsymbol{x} \in \mathbb{R}^N$, where $\boldsymbol{e}_1 = (1, 0, \ldots, 0)^\top$.

Next, we drive the formula for futures price under the multifactor Gaussian model.

Proposition 2.2. *The price function of the futures contract with maturity T is given by*

$$F(t, \boldsymbol{x}) = \exp\left(\boldsymbol{a}(t)^\top \boldsymbol{x} + \beta(t)\right), \tag{2.8}$$

for $(t, \boldsymbol{x}) \in [0, T) \times \mathbb{R}^N$, *where*

$$\boldsymbol{a}(t) = \exp\left(-(T-t)\boldsymbol{K}^\top\right)\boldsymbol{e}_1, \tag{2.9}$$

$$\beta(t) = \int_t^T (\boldsymbol{\mu} - \boldsymbol{\lambda})^\top \boldsymbol{a}(s) + \frac{1}{2}\operatorname{Tr}(\boldsymbol{\Sigma\Omega\Sigma}\boldsymbol{a}(s)\boldsymbol{a}(s)^\top)ds. \tag{2.10}$$

Proof. We substitute the ansatz solution (2.8) into PDE (2.7). The t-derivative is given by

$$\partial_t F(t, \boldsymbol{x}) = \left(\left(\frac{d\boldsymbol{a}(t)}{dt}\right)^\top \boldsymbol{x} + \frac{d\beta(t)}{dt}\right)\exp\left(\boldsymbol{a}(t)^\top \boldsymbol{x} + \beta(t)\right). \tag{2.11}$$

Then, the first and second derivatives satisfy

$$\nabla_{\boldsymbol{x}} F(t, \boldsymbol{x}) = \boldsymbol{a}(t) F(t, \boldsymbol{x}), \tag{2.12}$$

$$\nabla_{\boldsymbol{xx}} F(t, \boldsymbol{x}) = \boldsymbol{a}(t)\boldsymbol{a}(t)^\top F(t, \boldsymbol{x}). \tag{2.13}$$

By substituting (2.11), (2.12), and (2.13) into PDE (2.7), we obtain

$$\frac{d\boldsymbol{a}(t)}{dt} - \boldsymbol{K}^\top \boldsymbol{a}(t) = 0, \tag{2.14}$$

and

$$\frac{d\beta(t)}{dt} + (\boldsymbol{\mu} - \boldsymbol{\lambda})^\top \boldsymbol{a}(t) + \frac{1}{2}\operatorname{Tr}(\boldsymbol{\Sigma\Omega\Sigma}\boldsymbol{a}(t)\boldsymbol{a}(t)^\top) = 0. \tag{2.15}$$

The terminal conditions of $\boldsymbol{a}(t)$ and $\beta(t)$ are given by

$$\boldsymbol{a}(T) = \boldsymbol{e}_1,$$

$$\beta(T) = 0.$$

By direct substitution, the solutions to ODEs (2.14) and (2.15) are given by (2.9) and (2.10). □

Next, we consider the futures price process, denoted by F_t. Under the risk-neutral measure $\mathbb{Q}$, F_t is a martingale and satisfies the SDE

$$\begin{aligned}\frac{dF_t}{F_t} &= \frac{1}{F_t}\nabla_{\boldsymbol{x}}F(t,\boldsymbol{X}_t)^\top \boldsymbol{\Sigma}\boldsymbol{C}d\boldsymbol{Z}_t^{\mathbb{Q}} \\ &= \boldsymbol{a}(t)^\top \boldsymbol{\Sigma}\boldsymbol{C}d\boldsymbol{Z}_t^{\mathbb{Q}}. \end{aligned} \tag{2.16}$$

Next, using (2.4) and $\boldsymbol{\lambda} = \boldsymbol{\Sigma}\boldsymbol{C}\boldsymbol{\zeta}$, the $\mathbb{P}$-dynamics for F_t is given by

$$\frac{dF_t}{F_t} = \boldsymbol{a}(t)^\top \boldsymbol{\lambda}dt + \boldsymbol{a}(t)^\top \boldsymbol{\Sigma}\boldsymbol{C}d\boldsymbol{Z}_t^{\mathbb{P}}. \tag{2.17}$$

We observe from (2.17) that the drift term of the futures price SDE depends on the market prices of risk $\boldsymbol{\lambda}$ and $\boldsymbol{a}(t)$, which in turn depends on coefficient matrix $\boldsymbol{K}$ and maturity T. The diffusion term also depends on $\boldsymbol{a}(t)$, so the futures price has time-varying volatility even though the spot price process has constant volatility. This is intuitive given that futures price must converge to the spot price at expiration.

2.2 Two-Factor Models

In this section, we discuss the application of our framework to two well-known two-factor models: the Schwartz model and central tendency Ornstein–Uhlenbeck (CTOU) model. In both cases, Proposition 2.2 can be applied directly, and these models will be used for futures trading in Chapter 3.

2.2.1 *The Schwartz model*

The Schwartz model, introduced by Schwartz (1997), takes into account the stochastic convenience yield in commodity prices. The Schwartz model belongs to our multifactor Gaussian framework. Indeed, this amounts to setting the coefficients in SDE (2.16) and (2.17) to be

$$\boldsymbol{\mu} = \begin{bmatrix} \mu_1 - \sigma_1^2/2 \\ \kappa\alpha \end{bmatrix}, \quad \boldsymbol{K} = \begin{bmatrix} 0 & 1 \\ 0 & \kappa \end{bmatrix},$$

$$\boldsymbol{\Sigma} = \begin{bmatrix} \sigma_1 & 0 \\ 0 & \sigma_2 \end{bmatrix}, \quad \boldsymbol{C} = \begin{bmatrix} 1 & 0 \\ \rho & \sqrt{1-\rho^2} \end{bmatrix},$$

and

$$\boldsymbol{\lambda} = (\mu_1 - r, \lambda_2)^\top,$$

where the means μ_1, α, speed of mean reversion κ, interest rate r, the volatility parameters σ_1, σ_2 and market price of convenience yield risk λ_2 are constants and the instantaneous correlation ρ lies in $(-1,1)$.

In this case, the futures price $F^{(k)}(t,\boldsymbol{x})$ of maturity T_k was first obtained by Schwartz (1997). In our framework, we can also apply Proposition 2.2 and immediately arrive at the following formula:

$$F^{(k)}(t,\boldsymbol{x}) = \exp\left(\boldsymbol{a}^{(k)}(t)^\top \boldsymbol{x} + \beta^{(k)}(t)\right),$$

where

$$\boldsymbol{a}^{(k)}(t) = \left(1, \frac{e^{-(T_k-t)\kappa} - 1}{\kappa}\right)^\top$$

is a vector, and

$$\begin{aligned}\beta^{(k)}(t) = &\left(r - \hat{\alpha} - \frac{\rho\sigma_1\sigma_2}{\kappa} + \frac{\sigma_2^2}{2\kappa^2}\right)(T_k - t)\\ &+ \left(\kappa\hat{\alpha} + \rho\sigma_1\sigma_2 - \frac{\sigma_2^2}{\kappa}\right)\frac{1 - e^{-(T_k-t)\kappa}}{\kappa^2} + \frac{\sigma_2^2}{2}\frac{1 - e^{-2(T_k-t)\kappa}}{2\kappa^3},\end{aligned}$$

along with the constant

$$\hat{\alpha} = \alpha - \frac{\lambda_2}{\kappa}.$$

Therefore, the futures price function can be instantly computed.

2.2.2 *Numerical implementation and examples*

Next, we show numerical examples to illustrate futures price dynamic and term structure under the Schwartz model. The parameters are shown in the Table 2.1, which are from Table VI in Schwartz (1997). A Kalman filter approach to calibrate the Schwartz two-factor model can be found in Ewald *et al.* (2019). We note that in this setup the two factors $X^{(1)}$ and $X^{(2)}$ positively correlated with coefficient 0.845.

First, we simulate the sample paths for asset's spot and futures price in Figure 2.1. One defining characteristic of the Schwartz model is that the stochastic convenience yield is mean reverting. This can be observed in Figure 2.2. Moreover, the futures price and asset's spot price are seen to be highly correlated under the Schwartz model.

Figure 2.3 shows the term structure with different volatility parameter σ_1 and interest rate r. Figure 2.4 shows the term structure with different

Table 2.1. Parameters for the Schwartz model.

$X_0^{(1)}$	$X_0^{(2)}$	μ_1	κ	α	σ_1	σ_2
3	0.090	0.082	1.187	0.090	0.212	0.187
ρ	r	λ_2	T_1	T_2		
0.845	0.03	0.093	3/12	6/12		

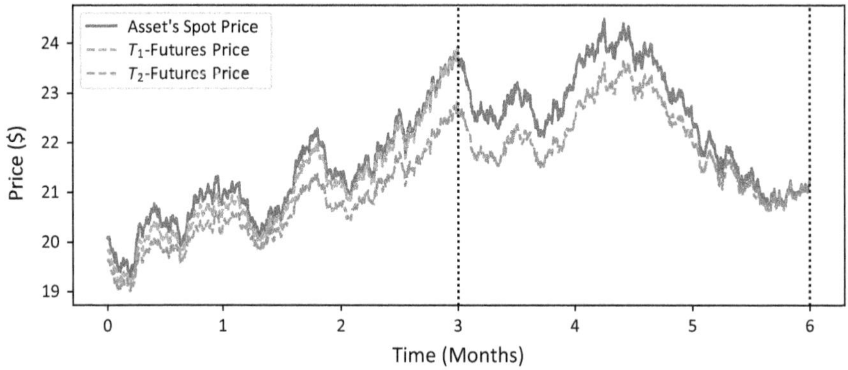

Fig. 2.1. This figure shows the sample paths of the asset's spot price and futures prices in the Schwartz model. The solid line represents the asset's spot price and the dashed lines represent futures prices. Parameters are shown in Table 2.1.

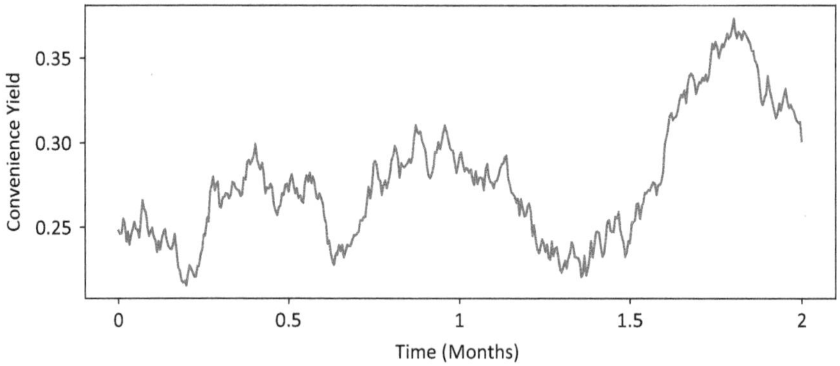

Fig. 2.2. The sample path of the stochastic convenience yield is shown. Under the Schwartz model, the convenience yield is mean-reverting. Parameters are shown in Table 2.1.

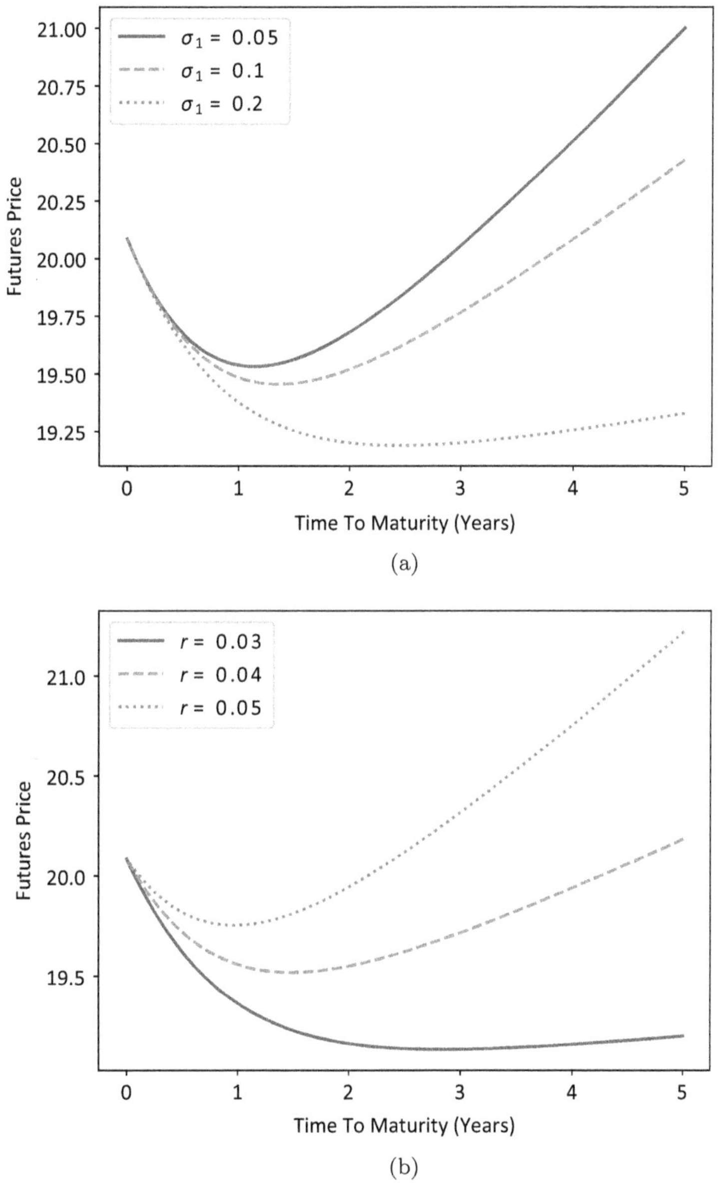

Fig. 2.3. The futures price as a function of (a) volatility σ_1 and (b) interest rate r in the Schwartz model. Parameters are shown in Table 2.1.

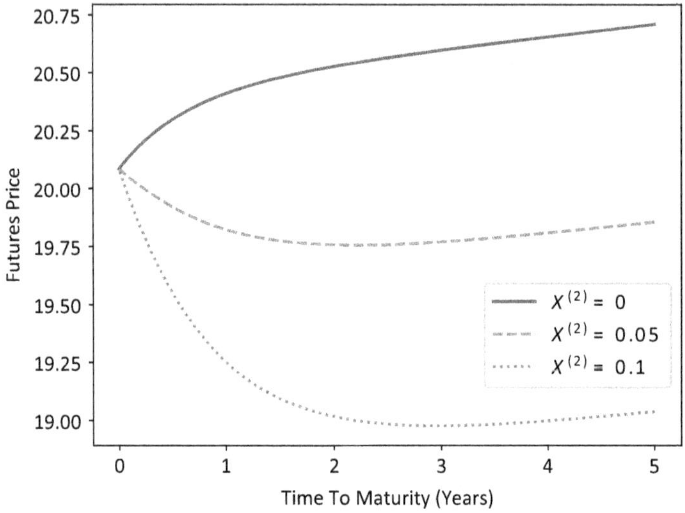

(a)

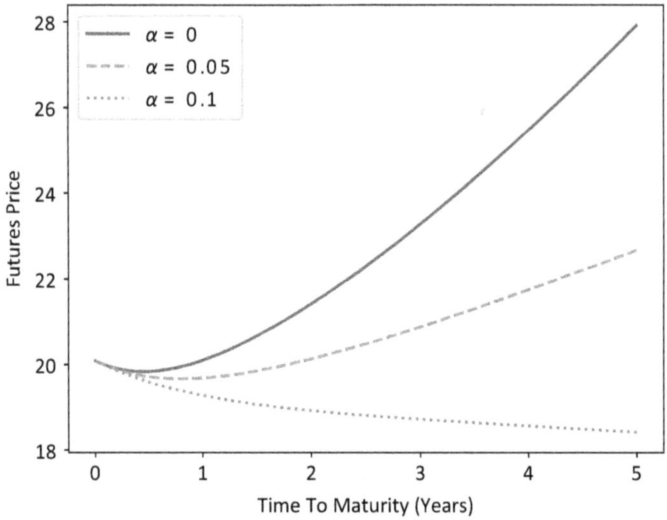

(a)

Fig. 2.4. The term structure with different initial convenience yield $X_0^{(2)}$ and long-term convenience yield mean α under the Schwartz model. Parameters are shown in Table 2.1.

initial convenience yield $X_0^{(2)}$ and long-term convenience yield mean α. Overall, the futures price decreases with respect to volatility parameter σ_1, initial convenience yield $X_0^{(2)}$ and long-term convenience yield mean α, but increases with respect to interest rate r. In most cases, the term structure exhibits a "smile" effect whereby the future price is decreasing at first and then increasing with respect to time-to-maturity.

Figure 2.5 shows the surface plot of 1-year futures price with respect to volatility parameter σ_1 and σ_2. We observe that the futures price is decreasing with respect to both volatility parameters. We also show the case with negative correlation ($\rho = -0.8$) in Figure 2.6. Interestingly, the futures price then is increasing with respect to both volatility parameters.

2.2.3 *The CTOU model*

CTOU model is studied by Mencia and Sentana (2013) for pricing VIX futures. Later, Leung and Yan (2018) analyze the futures portfolio optimization problem under this model. The CTOU process also belongs to

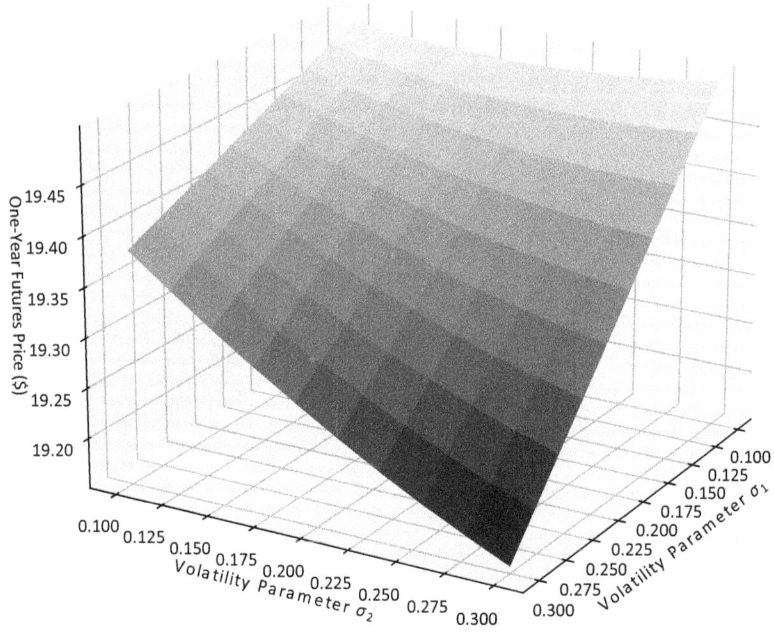

Fig. 2.5. The surface plot of 1-year futures price with respect to volatility parameter σ_1 and σ_2 under the Schwartz model. Parameters are shown in Table 2.1.

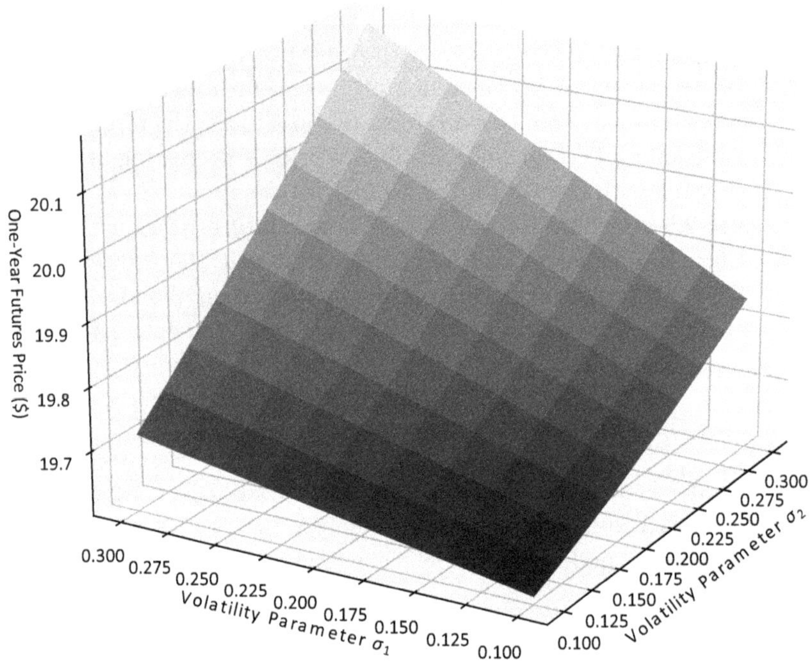

Fig. 2.6. The surface plot of 1-year futures price under the Schwartz model with respect to volatility parameter σ_1 and σ_2 with negative correlation parameter $\rho = -0.8$. Other parameters are shown in Table 2.1.

our multifactor framework. Indeed, this amounts to setting $N = 2$ and the coefficients in SDE (2.3) and (2.5) as

$$\boldsymbol{\mu} = \begin{bmatrix} \lambda_1 \\ \kappa_2\theta + \lambda_2 \end{bmatrix}, \quad \boldsymbol{K} = \begin{bmatrix} \kappa_1 & -\kappa_1 \\ 0 & \kappa_2 \end{bmatrix}, \tag{2.18}$$

$$\boldsymbol{\Sigma} = \begin{bmatrix} \sigma_1 & 0 \\ 0 & \sigma_2 \end{bmatrix}, \quad \boldsymbol{C} = \begin{bmatrix} 1 & 0 \\ \rho & \sqrt{1-\rho^2} \end{bmatrix},$$

and

$$\boldsymbol{\lambda} = (\lambda_1, \lambda_2)^\top, \tag{2.19}$$

where the mean θ, speeds of mean reversion $\{\kappa_1, \kappa_2\}$, the volatility parameters $\{\sigma_1, \sigma_2\}$, and adjusted market prices of risk $\{\lambda_1, \lambda_2\}$ are all constants. We note that in Mencia and Sentana (2013) and Leung and Yan (2018), the instantaneous correlation ρ is set to be 0.

Applying Proposition 2.2, the futures price $F^{(k)}(t, \boldsymbol{x})$ of maturity T_k in the CTOU model is given by

$$F^{(k)}(t, \boldsymbol{x}) = \exp\left(\boldsymbol{a}^{(k)}(t)^\top \boldsymbol{x} + \beta^{(k)}(t)\right),$$

where $\boldsymbol{a}^{(k)}(t)$ satisfies

$$\boldsymbol{a}^{(k)}(t) = \left(e^{-(T_k-t)\kappa_1}, \frac{\kappa_1}{\kappa_1 - \kappa_2}\left(e^{-(T_k-t)\kappa_2} - e^{-(T_k-t)\kappa_1}\right)\right)^\top,$$

and $\beta^{(k)}(t)$ follows:

$$\begin{aligned}
\beta^{(k)}(t) &= \theta - D_1(T_k - t)\theta + \frac{\sigma_1^2}{4\kappa_1}\left(1 - e^{-2\kappa_1(T_k-t)}\right) \\
&\quad + \rho\sigma_1\sigma_2\frac{\kappa_1}{\kappa_1 - \kappa_2}\left(\frac{1 - e^{-(\kappa_1+\kappa_2)(T_k-t)}}{\kappa_1 + \kappa_2} - \frac{1 - e^{-2\kappa_1(T_k-t)}}{2\kappa_1}\right) \\
&\quad + \frac{\sigma_2^2}{2}\left(\frac{\kappa_1}{\kappa_1 - \kappa_2}\right)^2\left(\frac{1 - e^{-2\kappa_2(T_k-t)}}{2\kappa_2} + \frac{1 - e^{-2\kappa_1(T_k-t)}}{2\kappa_1}\right. \\
&\quad \left. - 2\frac{1 - e^{-(\kappa_1+\kappa_2)(T_k-t)}}{\kappa_1 + \kappa_2}\right),
\end{aligned}$$

with

$$D_1(\tau) = \frac{\kappa_1}{\kappa_1 - \kappa_2}e^{-\kappa_2\tau} - \frac{\kappa_2}{\kappa_1 - \kappa_2}e^{-\kappa_1\tau}.$$

When we set $\rho = 0$, the result is consistent with Mencia and Sentana (2013) and Leung and Yan (2018).

2.2.4 *Numerical implementation and examples*

Next, we show numerical examples to illustrate futures price dynamic and term structure under the CTOU model. The parameters used for the subsequent figures are shown in Table 2.2, which is from the "full sample" in Table 4 of Mencia and Sentana (2013). We note that the initial values of $X_0^{(1)}$ and $X_0^{(2)}$ are the same as that of θ. The speed of mean reversion of underlying process $X^{(1)}$ is an order of magnitude greater than that of $X^{(2)}$. This means that $X^{(2)}$ tends to mean revert much slower than $X^{(1)}$. The volatility of $X^{(2)}$ is also smaller than the volatility of $X^{(1)}$.

We show the sample paths of the asset's spot price, futures price and stochastic mean in Figure 2.7. The sample path of the stochastic mean is presented in Figure 2.8, which is mean reverting. Unlike the

Table 2.2. Parameters for the CTOU model.

$X_0^{(1)}$	$X_0^{(2)}$	θ	κ_1	κ_2
3.019	3.019	3.019	5.827	0.300

σ_1	σ_2	ρ	ζ_1	ζ_2	T_1	T_2
1.037	0.446	−0.5	−0.010	2.242	1	2

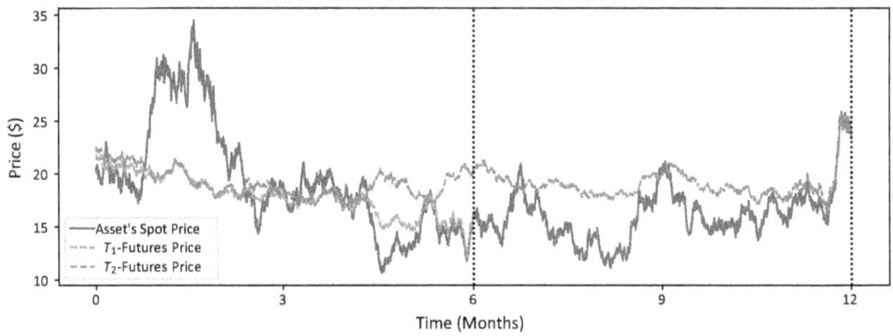

Fig. 2.7. The sample paths of the asset's spot price and futures prices under the CTOU model are simulated. The solid line represents the asset's spot price and the dashed lines represent futures prices. Parameters are shown in Table 2.2.

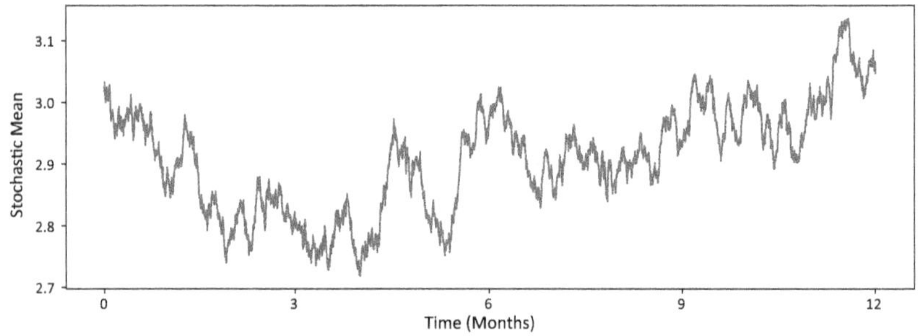

Fig. 2.8. This is the sample path of the stochastic mean corresponding to the simulated price processes in Figure 2.7. Under the CTOU model, the stochastic mean is a mean-reverting process. Parameters are shown in Table 2.2.

Schwartz model, the futures prices and asset's spot price are not highly correlated under the CTOU model as observed in these two figures. The T_1-futures price process runs for six months ends at maturity while the T_2-futures price continues until the end of the twelfth month.

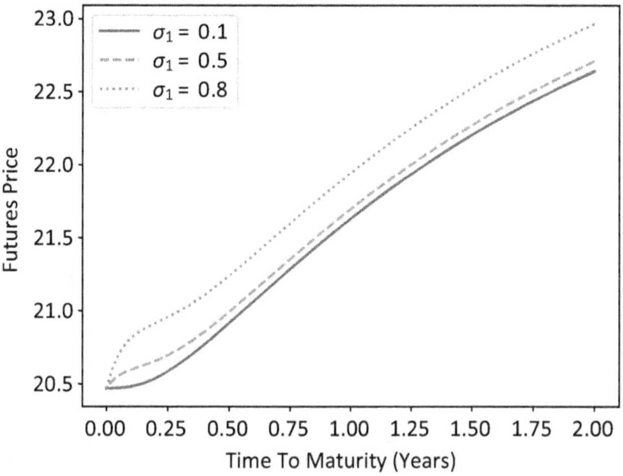

(a)

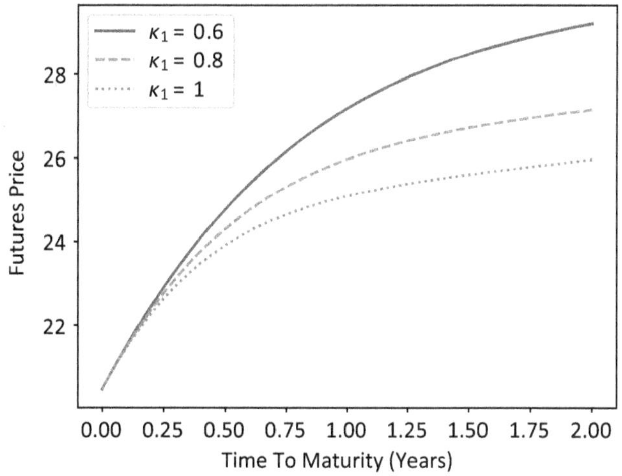

(b)

Fig. 2.9. The futures term structure under different (a) volatility levels and (b) mean reversion speed under the CTOU model. Parameters are shown in Table 2.2.

Over the course of a year, the simulated spot price has experienced much higher volatility than the futures prices.

Figure 2.9 shows the term structure with different volatility parameter σ_1 and mean reversion speed κ_1, while Figure 2.10 shows the term structure with different initial value for stochastic mean $X^{(2)}$ and correlation

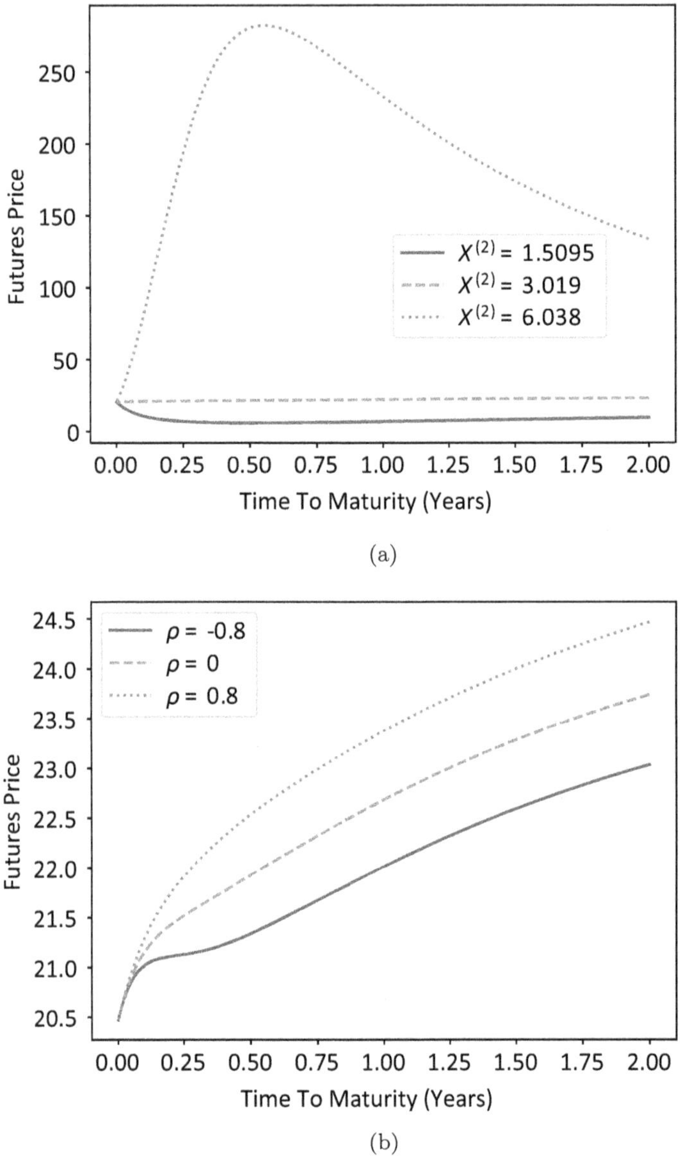

Fig. 2.10. The futures term structure under different (a) initial value of the stochastic mean $X^{(2)}$ and (b) correlation levels under the CTOU model. Parameters are shown in Table 2.2.

parameter ρ. Overall, the futures price increases with volatility parameter σ_1, initial value for stochastic mean $X^{(2)}$ and correlation parameter ρ, but decreases with respect to mean reversion speed κ_1.

In most cases, the futures price is increasing with respect to the maturity. As for the initial value for stochastic mean $X^{(2)}$, when it is smaller than its mean θ, the futures price is decreasing with respect to the maturity. However, when it is much higher than its mean θ, the futures price is increasing and then decreasing with respect to the maturity, since the futures price $X^{(1)}$ is driven by the instantaneous mean $X^{(2)}$ in the near future, and both futures price $X^{(1)}$ and it stochastic mean $X^{(2)}$ will drop back to its normal lever in the far future.

Figure 2.11 shows the surface plot of 1-year futures price with respect to volatility parameter σ_1 and correlation parameter ρ, while Figure 2.12 shows the surface plot of 1-year futures price with respect to volatility

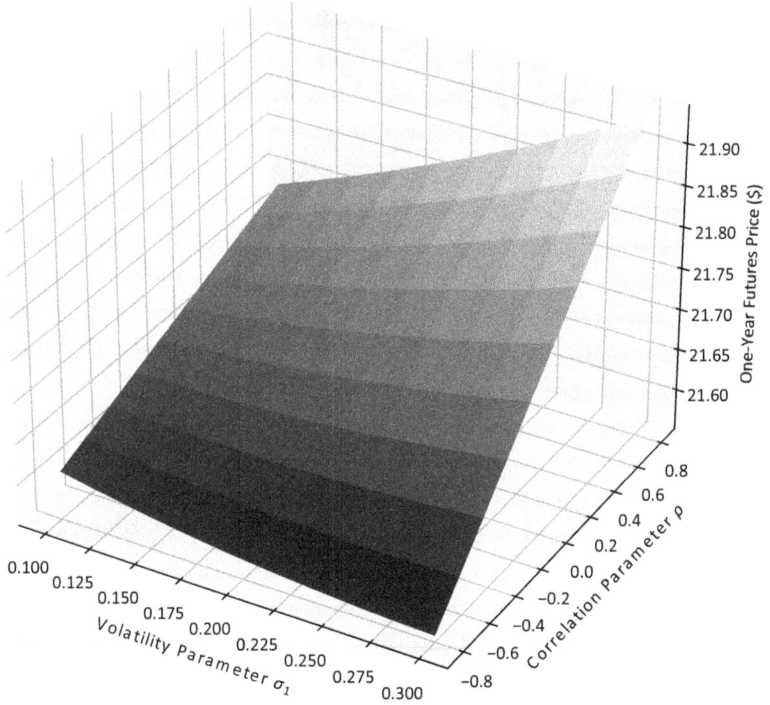

Fig. 2.11. The surface plot of 1-year futures price with respect to volatility parameter σ_1 and correlation parameter ρ under the CTOU model. Parameters are shown in Table 2.2.

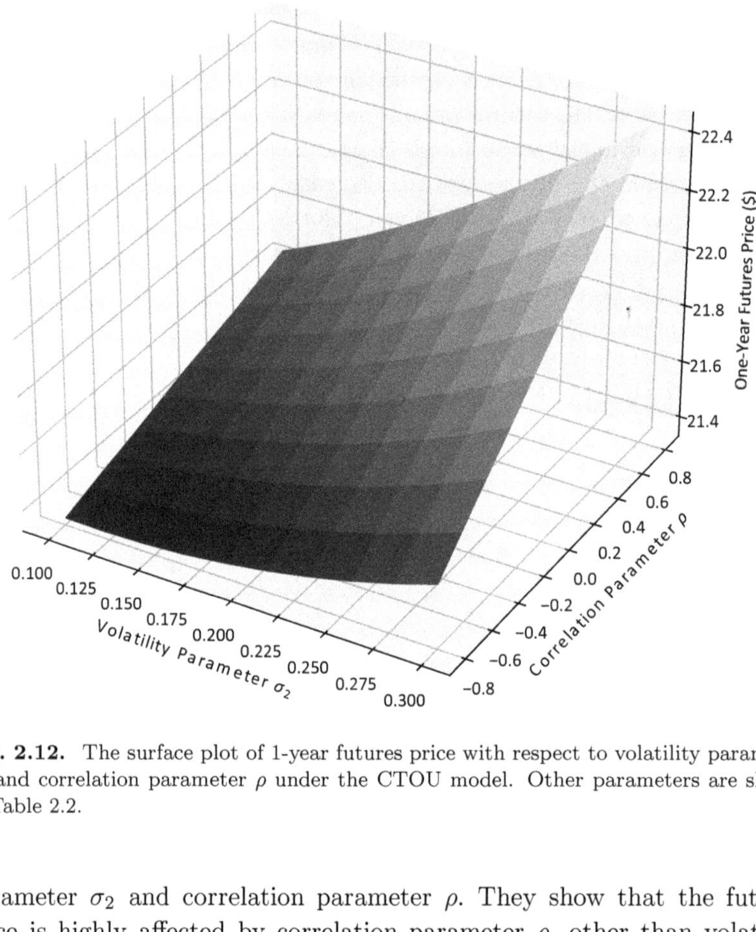

Fig. 2.12. The surface plot of 1-year futures price with respect to volatility parameter σ_2 and correlation parameter ρ under the CTOU model. Other parameters are shown in Table 2.2.

parameter σ_2 and correlation parameter ρ. They show that the futures price is highly affected by correlation parameter ρ, other than volatility parameters σ_1 and σ_2.

2.3 The Multiscale Central Tendency Ornstein–Uhlenbeck Model

In this section, we introduce the class of two-scale CTOU process. We also discuss the futures pricing and futures trading problem under this model.

2.3.1 *Model formulation*

Following the change of measure procedure described in Section 2.1, we denote the log-price of the underlying asset by $X_t^{(1)}$ and its evolution under the physical measure $\mathbb{P}$ is given by the system of SDEs:

$$
\begin{aligned}
dX_t^{(1)} &= \kappa\big(X_t^{(2)} + X_t^{(3)} - X_t^{(1)}\big)dt + \sigma_1 dZ_t^{\mathbb{P},1},\\
dX_t^{(2)} &= \frac{1}{\epsilon}\big(\alpha_2 - X_t^{(2)}\big)dt + \frac{1}{\sqrt{\epsilon}}\sigma_2\left(\rho_{12}dZ_t^{\mathbb{P},1} + \sqrt{1-\rho_{12}^2}dZ_t^{\mathbb{P},2}\right),\\
dX_t^{(3)} &= \delta\big(\alpha_3 - X_t^{(3)}\big)dt\\
&\quad + \sqrt{\delta}\sigma_3\left(\rho_{13}dZ_t^{\mathbb{P},1} + \rho_{23}dZ_t^{\mathbb{P},2} + \sqrt{1-\rho_{13}^2-\rho_{23}^2}dZ_t^{\mathbb{P},3}\right),
\end{aligned}
\tag{2.20}
$$

where $Z^{\mathbb{P},1}$, $Z^{\mathbb{P},2}$, and $Z^{\mathbb{P},3}$ are independent Brownian motions under the physical measure $\mathbb{P}$.

As such, the stochastic mean of log price $X^{(1)}$ is the sum of two factors $X_t^{(2)}$ and $X_t^{(3)}$. The first factor $X_t^{(2)}$ is a fast mean reverting diffusion process. We denote by $1/\epsilon$ the rate of mean reversion of this process, with $\epsilon > 0$ being a small parameter which corresponds to the time scale of this process. It is an ergodic process and its invariant distribution is independent of ϵ. This distribution is Gaussian with mean α_2 and variance $\sigma_2^2/2$. In addition, we assume that the correlation coefficient ρ_{12} is constant and $|\rho_{12}| < 1$.

The process $X_t^{(3)}$ is slowly varying, obtained by slowing down the diffusion process with a small parameter δ. We also assume the correlation coefficient ρ_{13} and ρ_{23} are constants, and they satisfy

$$\rho_{13}^2 + \rho_{23}^2 < 1.$$

Again, for futures pricing, we consider the risk-neutral pricing measure $\mathbb{Q}$. To that end, we connect the Brownian motions under the physical measure $\mathbb{P}$ and risk-neutral pricing measure $\mathbb{Q}$ via three risk premium factors. We denote the risk premium factors by ζ_i, for $i = 1, 2, 3$, which satisfy

$$dZ_t^{\mathbb{Q},i} = dZ_t^{\mathbb{P},i} + \zeta_i,$$

where $Z^{\mathbb{Q},1}$, $Z^{\mathbb{Q},2}$, and $Z^{\mathbb{Q},3}$ are independent Brownian motions under risk-neutral measure $\mathbb{Q}$.

Now, we introduce the combined market prices of volatility risk λ_i, for $i = 1, 2, 3$, defined by

$$
\begin{aligned}
\lambda_1 &= \zeta_1\sigma_1,\\
\lambda_2 &= \frac{1}{\sqrt{\epsilon}}\sigma_2\left(\zeta_1\rho_{12} + \zeta_2\sqrt{1-\rho_{12}^2}\right),\\
\lambda_3 &= \sqrt{\delta}\sigma_3\left(\zeta_1\rho_{13} + \zeta_2\rho_{23} + \zeta_3\sqrt{1-\rho_{13}^2-\rho_{23}^2}\right).
\end{aligned}
$$

Then, we write the evolution under the risk neutral measure as

$$
\begin{aligned}
dX_t^{(1)} &= \kappa\big(X_t^{(2)} + X_t^{(3)} - X_t^{(1)} - \lambda_1/\kappa\big)dt + \sigma_1 dZ_t^{\mathbb{Q},1}, \\
dX_t^{(2)} &= \frac{1}{\epsilon}\big(\alpha_2 - X_t^{(2)} - \epsilon\lambda_2\big)dt + \frac{1}{\sqrt{\epsilon}}\sigma_2\left(\rho_{12} dZ_t^{\mathbb{Q},1} + \sqrt{1-\rho_{12}^2}\, dZ_t^{\mathbb{Q},2}\right), \\
dX_t^{(3)} &= \delta\big(\alpha_3 - X_t^{(3)} - \lambda_3/\delta\big)dt \\
&\quad + \sqrt{\delta}\sigma_3\left(\rho_{13} dZ_t^{\mathbb{Q},1} + \rho_{23} dZ_t^{\mathbb{Q},2} + \sqrt{1-\rho_{13}^2-\rho_{23}^2}\, dZ_t^{\mathbb{Q},3}\right).
\end{aligned}
$$

This two-scale CTOU model belongs to our multifactor model framework discussed in the previous sections. Indeed, this amounts to setting the coefficients in SDE (2.3) and (2.5) to be

$$
\boldsymbol{\mu} = (0, \alpha_2/\epsilon, \delta\alpha_3)^\top, \quad \boldsymbol{\lambda} = (\lambda_1, \lambda_2, \lambda_3)^\top,
$$

$$
\boldsymbol{K} = \begin{bmatrix} \kappa & -\kappa & -\kappa \\ 0 & 1/\epsilon & 0 \\ 0 & 0 & \delta \end{bmatrix}, \quad \boldsymbol{\Sigma} = \begin{bmatrix} \sigma_1 & 0 & 0 \\ 0 & \sigma_2/\sqrt{\epsilon} & 0 \\ 0 & 0 & \sqrt{\delta}\sigma_3 \end{bmatrix},
$$

and

$$
\boldsymbol{C} = \begin{bmatrix} 1 & 0 & 0 \\ \rho_{12} & \sqrt{1-\rho_{12}^2} & 0 \\ \rho_{13} & \rho_{23} & \sqrt{1-\rho_{13}^2-\rho_{23}^2} \end{bmatrix}.
$$

Remark 2.3. If the stochastic mean of log price $X^{(1)}$ is only modeled by $X^{(2)}$ or $X^{(3)}$, instead of their sum, then it will reduce to the simpler CTOU model, which is discussed in Section 2.2.3. For example, if the log price $X^{(1)}$ evolves according to

$$
dX_t^{(1)} = \kappa\big(X_t^{(2)} - X_t^{(1)}\big)dt + \sigma_1 dZ_t^{\mathbb{P},1},
$$

where $X_t^{(2)}$ admits the process (2.20), we can just apply the theory in Section 2.2.3 by setting

$$
\boldsymbol{\mu} = \begin{bmatrix} 0 \\ \alpha_2/\epsilon \end{bmatrix}, \quad \boldsymbol{K} = \begin{bmatrix} \kappa & -\kappa \\ 0 & 1/\epsilon \end{bmatrix},
$$

$$
\boldsymbol{\Sigma} = \begin{bmatrix} \sigma_1 & 0 \\ 0 & \sigma_2/\sqrt{\epsilon} \end{bmatrix}, \quad \boldsymbol{C} = \begin{bmatrix} 1 & 0 \\ \rho_{12} & \sqrt{1-\rho_{12}^2} \end{bmatrix},
$$

and

$$\boldsymbol{\lambda} = (\lambda_1, \lambda_2)^\top,$$

in (2.18) and (2.19).

2.3.2 *Futures pricing*

First, we consider three futures contracts $F^{(1)}$, $F^{(2)}$, and $F^{(3)}$, written on this underlying, whose maturity are T_1, T_2, and T_3, respectively. For convenience, we define the following functions:

$$f_k(p,t) := e^{-(T_k-t)p},$$

$$g_k(p,t) := \frac{1-e^{-(T_k-t)p}}{p}.$$

Then, applying Proposition 2.2, the futures price $F^{(k)}(t,\boldsymbol{x})$ of maturity T_k under this multiscale CTOU model satisfies

$$F^{(k)}(t,\boldsymbol{x}) = \exp\left(\boldsymbol{a}^{(k)}(t)^\top \boldsymbol{x} + \beta^{(k)}(t)\right), \tag{2.21}$$

where $\boldsymbol{a}^{(k)}(t)$ satisfies

$$\begin{aligned}\boldsymbol{a}^{(k)}(t) = \Bigg(& f_k(\kappa,t), \frac{\epsilon\kappa}{1-\epsilon\kappa}\left(f_k(\kappa,t) - f_k(1/\epsilon,t)\right),\\ & \frac{\kappa}{\delta-\kappa}\left(f_k(\kappa,t) - f_k(\delta,t)\right)\Bigg)^\top,\end{aligned}$$

and $\beta^{(k)}(t)$ is given by

$$\begin{aligned}\beta^{(k)}(t) &= \int_t^{T_k} (\boldsymbol{\mu}-\boldsymbol{\lambda})^\top \boldsymbol{a}^{(k)}(s) + \frac{1}{2}\operatorname{Tr}(\boldsymbol{\Sigma\Omega\Sigma} \boldsymbol{a}^{(k)}(s)\boldsymbol{a}^{(k)}(s)^\top) ds\\ &= \left(-\lambda_1 + \frac{\epsilon\kappa}{1-\epsilon\kappa}(\alpha_2/\epsilon - \lambda_2) + \frac{\kappa}{\delta-\kappa}(\delta\alpha_3 - \lambda_3)\right) g_k(\kappa,s)\\ &\quad - \frac{\epsilon\kappa(\alpha_2/\epsilon-\lambda_2)}{1-\epsilon\kappa} g_k(1/\epsilon,s) - \frac{\kappa(\delta\alpha_3-\lambda_3)}{\delta-\kappa} g_k(\delta,s)\\ &\quad + \frac{1}{2}\Bigg(1 + \left(\frac{\epsilon\kappa}{1-\epsilon\kappa}\right)^2 + \left(\frac{\kappa}{\delta-\kappa}\right)^2 + \frac{2\rho_{12}\epsilon\kappa}{1-\epsilon\kappa} + \frac{2\rho_{13}\kappa}{\delta-\kappa}\\ &\quad + \frac{2\left(\rho_{12}\rho_{13} + \rho_{23}\sqrt{1-\rho_{12}^2}\right)\epsilon\kappa^2}{(1-\epsilon\kappa)(\delta-\kappa)}\Bigg) g_k(2\kappa,t)\end{aligned}$$

$$
\begin{aligned}
&+\frac{1}{2}\left(\frac{\epsilon\kappa}{1-\epsilon\kappa}\right)^2 g_k(2/\epsilon,s)+\frac{1}{2}\left(\frac{\kappa}{\delta-\kappa}\right)^2 g_k(2\delta,s)\\
&-\left(\left(\frac{\epsilon\kappa}{1-\epsilon\kappa}\right)^2+\rho_{12}\frac{\epsilon\kappa}{1-\epsilon\kappa}+\left(\rho_{12}\rho_{13}+\rho_{23}\sqrt{1-\rho_{12}^2}\right)\right.\\
&\qquad\left.\frac{\epsilon\kappa^2}{(1-\epsilon\kappa)(\delta-\kappa)}\right)g_k(1/\epsilon+\kappa,s)\\
&-\left(\left(\frac{\kappa}{\delta-\kappa}\right)^2+\rho_{13}\frac{\kappa}{\delta-\kappa}+\left(\rho_{12}\rho_{13}+\rho_{23}\sqrt{1-\rho_{12}^2}\right)\right.\\
&\qquad\left.\frac{\epsilon\kappa^2}{(1-\epsilon\kappa)(\delta-\kappa)}\right)g_k(\delta+\kappa,s)\\
&+\left(\rho_{12}\rho_{13}+\rho_{23}\sqrt{1-\rho_{12}^2}\right)\frac{\epsilon\kappa^2}{(1-\epsilon\kappa)(\delta-\kappa)}g_k(1/\epsilon+\delta,s).
\end{aligned}
$$

2.4 Numerical Illustration

We begin with numerical examples for the multiscale CTOU model discussed in Section 2.3. With the closed-form expressions obtained, we now implement the futures prices numerically, using the parameters in Table 2.3. Primarily, we let ϵ and δ be small parameters and we consider trading three futures with maturities $T_1 = 3/12$ year, $T_2 = 6/12$ year, and $T_3 = 9/12$ year.

In Figure 2.13, we plot the simulated paths and 95% confidence intervals for three factors in Figures 2.13(a) and 2.13(b). As shown in Figure 2.13(b), the 95% confidence interval of the slow-varying factor $X^{(3)}$ is much narrower than the one for fast-varying factor $X^{(2)}$. In Figure 2.13(c), we plot the spot price in the solid lines and futures prices in the dashed lines.

In Figure 2.14, we generate the term structure of futures prices with different volatility parameter σ_1 and initial log-price $X_0^{(1)}$, while we show the

Table 2.3. Parameters for the multiscale CTOU model.

$X_0^{(1)}$	$X_0^{(2)}$	$X_0^{(3)}$	α_2	α_3	κ	ϵ	δ	σ_1	σ_2	σ_3
1	0.5	0.5	0.5	0.5	5	0.1	0.1	0.5	0.3	0.3

ρ_{12}	ρ_{13}	ρ_{23}	λ_1	λ_2	λ_3	T_1	T_2	T_3
0	0	0	0.02	0.02	0.02	3/12	6/12	9/12

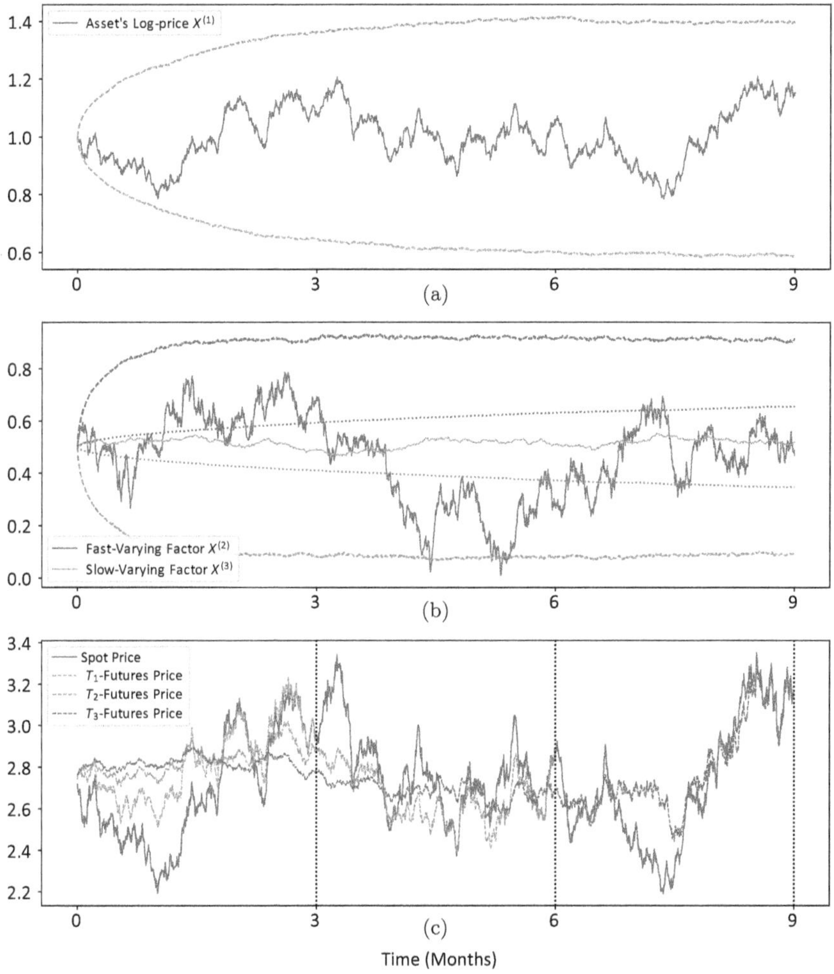

Fig. 2.13. (a) Simulated paths for asset's log-price $X^{(1)}$ under the multiscale CTOU model. Dashed curves represent 95% confidence intervals. (b) Simulated path for the fast-varying factor $X^{(2)}$ and slow-varying factor $X^{(2)}$ for the MCTOU model. Dashed curves represent 95% confidence intervals for $X^{(2)}$ and dotted curves represent 95% confidence intervals for $X^{(3)}$. (c) Simulated paths for asset's spot price and futures prices under the MCTOU model. Parameters are shown in Table 2.3.

term structure of futures prices with different initial value for fast-varying factor $X^{(2)}$ and initial value for slow-varying factor $X^{(3)}$ in Figure 2.15. Overall, the futures price increases with respect to these four parameters. We shall see that with different initial log-price $X_0^{(1)}$ and initial value for

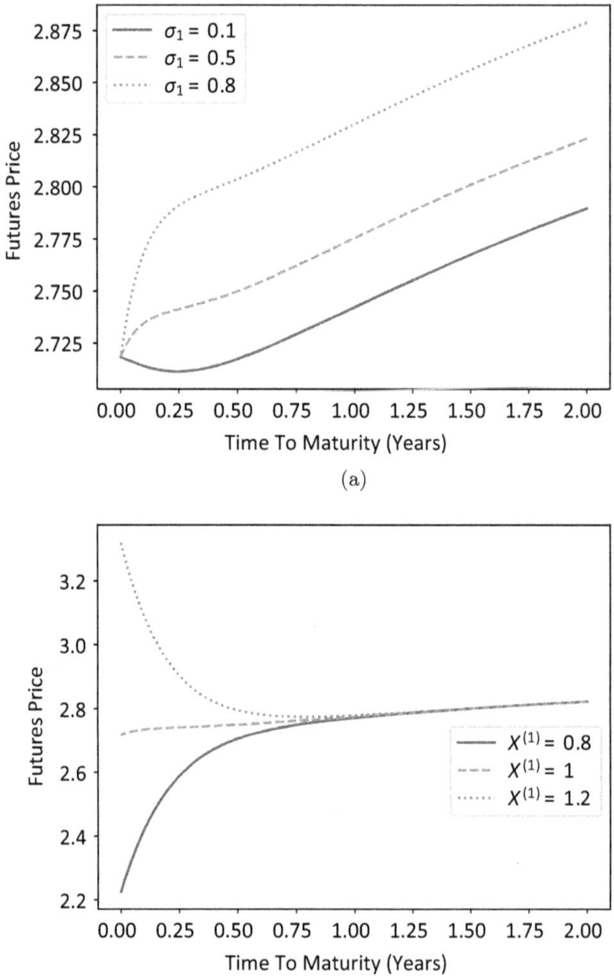

(a)

(b)

Fig. 2.14. The term structure with different volatility parameter σ_1 and initial log-price $X_0^{(1)}$ for the MCTOU model. Parameters are shown in Table 2.3.

fast-varying factor $X^{(2)}$, it may present different term structure in the near future, but it will converge in the far future, due to the mean reversion. However, we do not observe similar property for slow-varying factor $X^{(3)}$, since its mean reversion speed is too slow.

Figure 2.16 shows the surface plot of 1-year futures price with respect to correlation parameter ρ_{12} and ρ_{13}. Interestingly, when ρ_{12} is very positive,

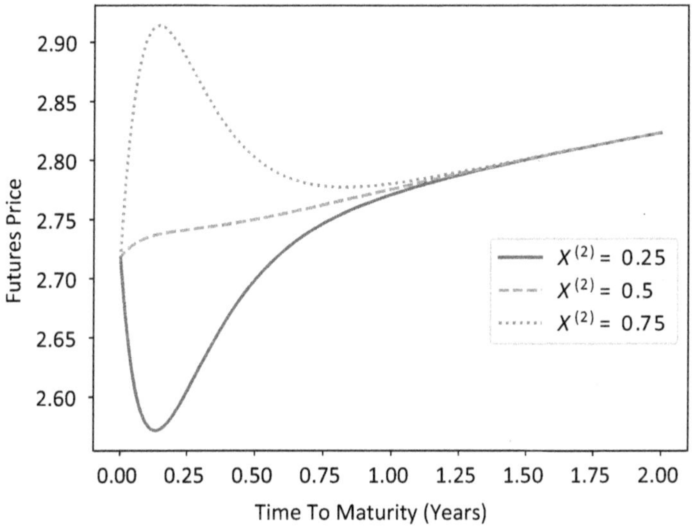

(a)

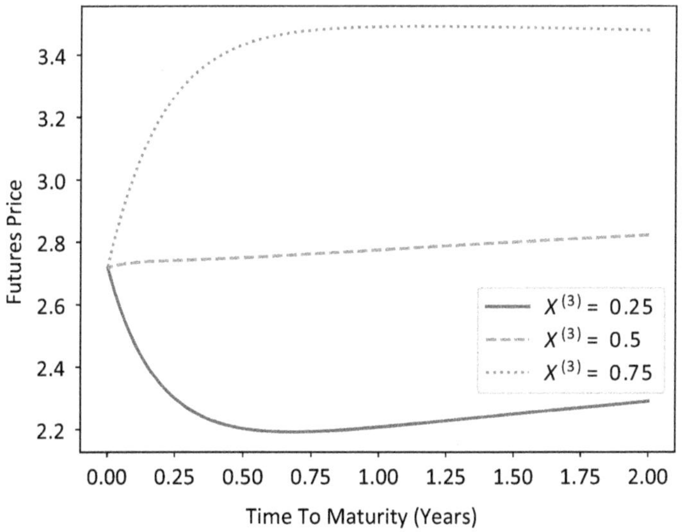

(b)

Fig. 2.15. The term structure with different initial value for fast-varying factor $X^{(2)}$ and initial value for slow-varying factor $X^{(3)}$ for the MCTOU model. Parameters are shown in Table 2.3.

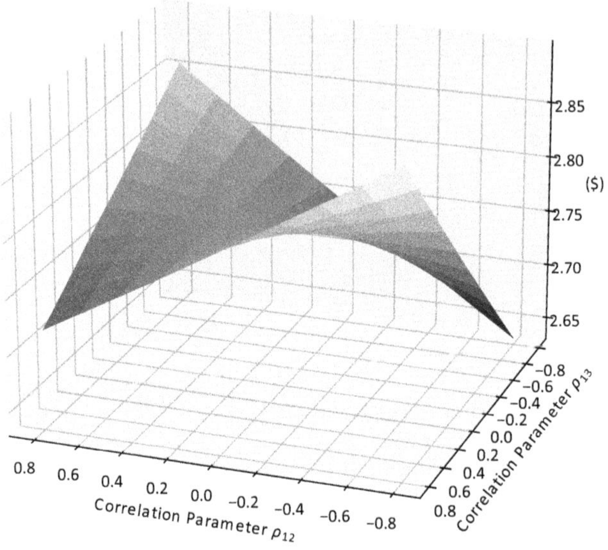

(a)

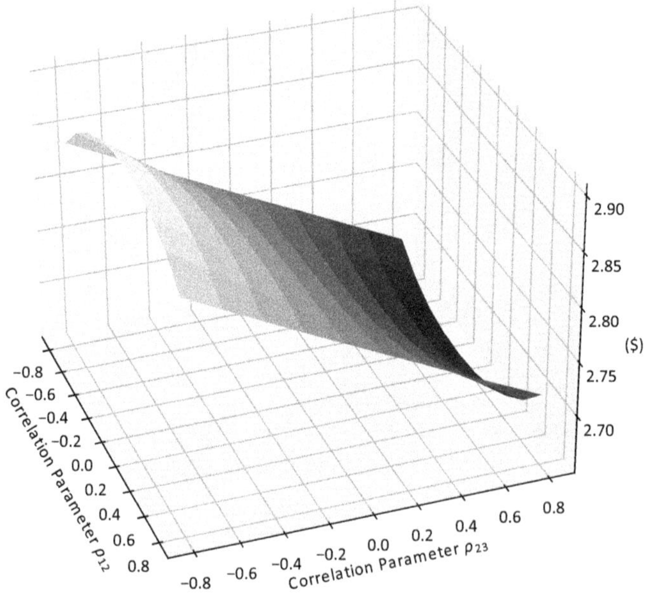

(b)

Fig. 2.16. The surface plot of 1-year futures price with respect to correlation parameters ρ_{12} and ρ_{13} in the MCTOU model. Parameters are shown in Table 2.3.

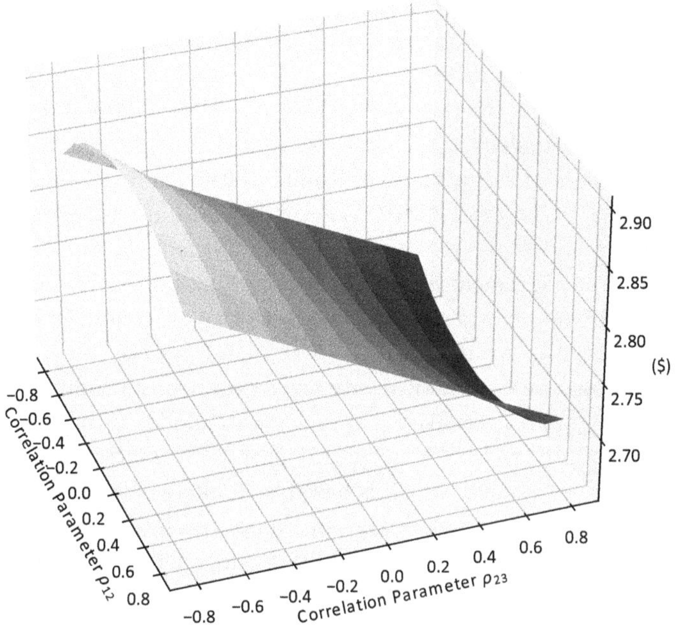

Fig. 2.17. The surface plot of 1-year futures price with respect to correlation parameters ρ_{12} and ρ_{23} for the MCTOU model. Parameters are shown in Table 2.3.

the futures price is decreasing with respect to ρ_{13}. In contrast, when ρ_{12} is very negative, the futures price is increasing with respect to ρ_{13}. As a consequence, the futures prices tend to be higher when the two correlation parameters take values of opposite signs.

Figure 2.17 shows the surface plot of 1-year futures price with respect to correlation parameter ρ_{12} and ρ_{23}. Overall, the futures price is decreasing with respect to ρ_{23}. However, its convexity with respect to ρ_{12} changes as ρ_{23} varies. As the figrue shows, the futures price peaks when ρ_{23} is most negative while ρ_{12} is close to 0. The lowest future price is found where ρ_{23} is close to 1 and ρ_{23} is near 0.

2.5 Conclusion

In this chapter, we have extended the study of pricing commodity futures market under two-factor models to a multifactor model. Closed-form expressions for the futures prices are obtained. Using these, we illustrated the optimal strategies. Intuitively, it should be more valuable to the investor to access a larger set of securities, and this intuition is confirmed quantitatively through the certainty equivalents associated with the optimal futures portfolios.

One direction for further research is model estimation. In fact, one advantage of the multifactor model, as argued by Cortazar and Naranjo (2006), is the ease of estimation via Kalman filter. In other related studies (Schwartz, 1997; Guo *et al.*, 2019; Vega, 2018), Kalman filtering and other calibration methodologies can handle multifactor models with hidden state variables and measurement errors. Inclusion of measurement errors is necessary since the number of available market prices is generally higher than the number of state variables that need to be estimated.

It would also be of practical interest to connect or combine the multifactor model with alternative approaches to model futures prices. For example, Barndorff-Nielsen *et al.* (2015) propose a multivariate model for commodity forward curves which is based on multivariate ambit fields, and Chen *et al.* (2023) discuss a dynamic copula approach to capture the dependence structure of various US commodity futures across different sectors.

Chapter 3

Dynamic Futures Portfolios Under a Multifactor Gaussian Framework

3.1 Introduction

In this chapter, we discuss the optimal dynamic futures trading strategies under a general Gaussian framework discussed in Chapter 2. Our framework incorporates a wide array of multifactor models, like the Schwartz (1997) model and central tendency Ornstein–Uhlenbeck (CTOU) model. With the price dynamics of the futures contracts derived in the previous chapter, we now discuss how to construct and optimize a dynamic futures portfolio.

The optimal futures trading strategy is determined by solving a stochastic control problem, whose objective is to maximize the expected utility from trading wealth. By analyzing and solving the associated Hamilton–Jacobi–Bellman (HJB) equations, we present the value function and optimal trading strategies explicitly.

In order to quantify the value of the futures trading opportunity, we define the portfolio manager's certainty equivalent. Intuitively, it should be more beneficial to be able to trade a larger set of securities. Using certainty equivalent, we quantify the value of trading different sets of futures, and show that the highest certainty equivalent is achieved from trading all available contracts. On the other hand, it is surprising that the certainty equivalent does not depend on the current spot and futures prices. We apply our stochastic framework to the Schwartz model and CTOU model. In addition, we introduce a new multiscale CTOU model that is driven by a fast and slow mean-reverting process. We provide the numerical examples to examine model parameters for our new model.

The continuous-time stochastic control approach for portfolio optimization dates back to Merton (1971), but much less has been done for dynamic futures portfolios. The utility maximization approach is used to derive dynamic futures trading strategies under two-factor models in Leung and Yan (2018) and Leung and Yan (2019). A regime-switching framework for dynamic futures trading can be found in Leung and Zhou (2019).

There are a few alternative mathematical approaches and applications of futures trading. Deng *et al.* (2020) study the trading and hedging of bitcoin futures under a mean–variance framework. As an alternative approach for capturing futures and spot price dynamics, the stochastic basis model Angoshtari and Leung (2019b, 2020) directly models the difference between the futures and underlying asset prices, and solve for the optimal trading strategies through utility maximization. In other applications, Benth and Karlsen (2005) study the Merton portfolio optimization problem under the Schwartz mean-reverting model, and Cartea and Jaimungal (2016) and Guijarro-Ordonez (2019) apply stochastic control methods to trading cointegrated securities in multifactor models. In practice, dynamic futures portfolios are also commonly used to track a commodity index Leung and Ward (2015). In comparison to these studies, we have extended the investigation of optimal trading in commodity futures market under two-factor models to a multi-factor model. Closed-form expressions for the optimal controls and for the value function are obtained. Using these formulae, the optimal futures positions can be instantly computed, as we illustrate in this chapter.

The rest of this chapter is structured as follows. We describe the general market framework and futures dynamics in Section 2.1. The futures portfolio optimization is discussed in Section 3.2. Then, we apply our framework to Schwartz model and CTOU process in Section 3.3. In addition, we introduce the multiscale CTOU model and apply our framework in Section 3.4. Lastly, we provide numerical analysis in Section 3.5. Concluding remarks are provided in Section 3.6.

3.2 Optimal Dynamic Futures Portfolio

We recall from Chapter 2 the multifactor market model. the N-dimensional vector $\boldsymbol{X}_t = (X_t^{(1)}, \ldots, X_t^{(N)})^\top$, where $X^{(1)}$ is the log-price of the underlying asset and $(X^{(2)}, \ldots, X^{(N)})$ are observable stochastic factors

Next, we consider a portfolio of M contracts of different maturities available to trade under N-factor model. Since there are N sources of randomness in the model, trading N or more than N futures will result in hedging away all the risk. Henceforth, we set $M \leq N$.

We will denote by $F^{(k)}(t, \boldsymbol{x})$ as the price function of futures contract with maturity T_k, with $T_1 < \cdots < T_M$, and by $F_t^{(k)}$ as the stochastic process for this contract, for $k = 1, \ldots, M$. Recall from (2.17) that the T_k-futures price process satisfies

$$\begin{aligned}\frac{dF_t^{(k)}}{F_t^{(k)}} &= \boldsymbol{a}^{(k)}(t)^\top \boldsymbol{\lambda} dt + \boldsymbol{a}^{(k)}(t)^\top \boldsymbol{\Sigma} \boldsymbol{C} d\boldsymbol{Z}_t^{\mathbb{P}} \\ &\equiv \mu_F^{(k)}(t)dt + \boldsymbol{\sigma}_F^{(k)}(t)^\top d\boldsymbol{Z}_t^{\mathbb{P}},\end{aligned}$$

where we have defined

$$\mu_F^{(k)}(t) \equiv \boldsymbol{a}^{(k)}(t)^\top \boldsymbol{\lambda}, \tag{3.1}$$

$$\boldsymbol{\sigma}_F^{(k)}(t) \equiv \boldsymbol{C}^\top \boldsymbol{\Sigma}^\top \boldsymbol{a}^{(k)}(t). \tag{3.2}$$

Then, in matrix form, the system of futures dynamics is given by the set of SDE:

$$d\boldsymbol{F}_t = \boldsymbol{\mu}_{\boldsymbol{F}}(t)dt + \boldsymbol{\Sigma}_{\boldsymbol{F}}(t)d\boldsymbol{Z}_t^{\mathbb{P}},$$

where

$$d\boldsymbol{F}_t = \left(\frac{dF_t^{(1)}}{F_t^{(1)}}, \ldots, \frac{dF_t^{(M)}}{F_t^{(M)}}\right)^\top,$$

$$\boldsymbol{\mu}_{\boldsymbol{F}}(t) = \left(\mu_F^{(1)}(t), \ldots, \mu_F^{(M)}(t)\right)^\top, \tag{3.3}$$

$$\boldsymbol{\Sigma}_{\boldsymbol{F}}(t) = \left(\boldsymbol{\sigma}_F^{(1)}(t), \ldots, \boldsymbol{\sigma}_F^{(M)}(t)\right)^\top. \tag{3.4}$$

Here, we assume there be no redundant futures contract in the portfolio, which means any futures contract could not be replicated by other $M - 1$ futures contracts. To that end, we require that $\text{rank}(\boldsymbol{\Sigma}_{\boldsymbol{F}}) = M$. Since $\text{rank}(\boldsymbol{\Sigma}_{\boldsymbol{F}}) \neq M$ if $M > N$, the rank condition effectively excludes the case with more contracts than stochastic factors, as desired.

Next, we consider the futures trading problem for the investor. Let us denote the dynamic strategy by the process

$$\boldsymbol{\pi}_t = \left(\pi_t^{(1)}, \ldots, \pi_t^{(M)}\right)^\top,$$

for $t \in [0, T]$, where the element $\pi_t^{(k)}$ denotes the amount of money invested in kth futures contract at time t. In addition, we assume the interest rate to be zero for simplicity.

Then, for any admissible strategy $\boldsymbol{\pi}$, the wealth process is

$$\begin{aligned} dW_t^{\pi} &= \sum_{k=1}^{M} \pi_t^{(k)} \frac{dF_t^{(k)}}{F_t^{(k)}} \\ &= \boldsymbol{\pi}_t^{\top} \boldsymbol{\mu}_{\boldsymbol{F}}(t)dt + \boldsymbol{\pi}_t^{\top} \boldsymbol{\Sigma}_{\boldsymbol{F}}(t) d\boldsymbol{Z}_t^{\mathbb{P}}. \end{aligned} \tag{3.5}$$

We note that the wealth process is only determined by the strategy $\boldsymbol{\pi}_t$ and it is not affected by factors variable $\boldsymbol{X}$ and futures prices $\boldsymbol{F}$.

We consider a utility maximization approach to determine the optimal strategy. The investor's risk preference is described by the exponential utility

$$U(w) = -\exp(-\gamma w),$$

where $\gamma > 0$ is the risk aversion parameter. The investor fixes a finite trading horizon $0 < \widetilde{T} \leq T_1$, which means that $\widetilde{T}$ has to be less than or equal to the maturity of the earliest expiring contract.

A strategy $\boldsymbol{\pi}$ is said to be admissible if $\boldsymbol{\pi}$ is real-valued progressively measurable and satisfies the Novikov (1972) condition

$$\mathbb{E}^{\mathbb{P}}\left[\exp\left(\int_t^{\widetilde{T}} \frac{\gamma^2}{2} \boldsymbol{\pi}_s^{\top} \boldsymbol{\Sigma}_{\boldsymbol{F}}(s) \boldsymbol{\Sigma}_{\boldsymbol{F}}^{\top}(s) \boldsymbol{\pi}_s ds\right)\right] < \infty. \tag{3.6}$$

The investor seeks to maximize the expected utility of wealth at $\widetilde{T}$ by solving the stochastic control problem

$$V(t, w) = \sup_{\boldsymbol{\pi} \in \mathcal{A}_t} \mathbb{E}^{\mathbb{P}}[U(W_{\widetilde{T}}^{\boldsymbol{\pi}}) | W_t = w], \tag{3.7}$$

where $\mathcal{A}_t$ denotes the set of admissible controls at the initial time t. Since the wealth SDE (3.5) does not depend on the factors variable $\boldsymbol{X}$ and futures prices $\boldsymbol{F}$, the value function does not depend on them either.

Following the standard verification approach to dynamic programming (Fleming and Soner, 1993; Ross, 2008; Oksendal, 2014), we let $C^{1,2}$ denote the set of all continuous functions $f(t, x)$ that are continuously differentiable in t and twice continuously differentiable in x. Then, we assume the existence of a sufficiently smooth candidate solution $u(t,w) \in C^{1,2}$, which will later be shown to be equal to the value function V in (3.7).

To facilitate presentation, we define

$$\mathcal{L}^{\pi} \cdot = \boldsymbol{\pi}_t^{\top} \boldsymbol{\mu}_{\boldsymbol{F}}(t)\partial_w \cdot + \frac{1}{2}\boldsymbol{\pi}_t^{\top} \boldsymbol{\Sigma}_{\boldsymbol{F}}(t)\boldsymbol{\Sigma}_{\boldsymbol{F}}^{\top}(t)\boldsymbol{\pi}_t \partial_{ww} \cdot$$

Then, the candidate value function $u(t,w)$ and optimal trading strategy π^* is found from the HJB equation

$$\partial_t u + \sup_{\boldsymbol{\pi}} \mathcal{L}^{\boldsymbol{\pi}} u = 0, \tag{3.8}$$

for $(t,w) \in [0,\widetilde{T}) \times \mathbb{R}$, along with the terminal condition

$$u(T,w) = -e^{-\gamma w}, \quad \text{for } w \in \mathbb{R}.$$

We summarize the solution of the futures portfolio optimization problem in the following theorem.

Theorem 3.1.

(1) *Define*

$$\Lambda^2(t) = \boldsymbol{\mu}_{\boldsymbol{F}}(t)^{\top} \left(\boldsymbol{\Sigma}_{\boldsymbol{F}}(t)\boldsymbol{\Sigma}_{\boldsymbol{F}}(t)^{\top}\right)^{-1} \boldsymbol{\mu}_{\boldsymbol{F}}(t). \tag{3.9}$$

The unique solution to the HJB equation (3.8) *is*

$$u(t,w) = -\exp\left(-\gamma w - \frac{1}{2}\int_t^{\widetilde{T}} \Lambda^2(s)ds\right). \tag{3.10}$$

(2) *The optimal strategy is given by*

$$\boldsymbol{\pi}^*(t) = \frac{1}{\gamma}\left(\boldsymbol{\Sigma}_{\boldsymbol{F}}(t)\boldsymbol{\Sigma}_{\boldsymbol{F}}^{\top}(t)\right)^{-1} \boldsymbol{\mu}_{\boldsymbol{F}}(t). \tag{3.11}$$

Proof. We will first use the ansatz

$$u(t,w) = -e^{-\gamma w}h(t),$$

to factor out w. Using the relations

$$\partial_t u = -e^{-\gamma w}\partial_t h(t),$$
$$\partial_w u = \gamma e^{-\gamma w}h(t),$$
$$\partial_{ww} u = -\gamma^2 e^{-\gamma w}h(t),$$

the PDE (3.8) becomes

$$-\frac{d}{dt}h(t) + \sup_{\boldsymbol{\pi}_t}\left[\gamma\boldsymbol{\pi}_t^\top\boldsymbol{\mu}_{\boldsymbol{F}}(t)h - \frac{1}{2}\gamma^2\boldsymbol{\pi}_t^\top\boldsymbol{\Sigma}_{\boldsymbol{F}}(t)\boldsymbol{\Sigma}_{\boldsymbol{F}}^\top(t)\boldsymbol{\pi}_t h\right] = 0, \quad (3.12)$$

with terminal condition $h(\widetilde{T}) = 1$. From the first-order condition, which is obtained from differentiating the terms inside the supremum with respect to $\boldsymbol{\pi}_t$ and setting the equation to zero, we have

$$\gamma\boldsymbol{\mu}_{\boldsymbol{F}}(t) - \gamma^2\boldsymbol{\Sigma}_{\boldsymbol{F}}(t)\boldsymbol{\Sigma}_{\boldsymbol{F}}^\top(t)\boldsymbol{\pi}_t = 0,$$

Recall that $\text{rank}(\boldsymbol{\Sigma}_{\boldsymbol{F}}(t)) = M$. Then, $\boldsymbol{\Sigma}_{\boldsymbol{F}}(t)\boldsymbol{\Sigma}_{\boldsymbol{F}}^\top(t)$ is an $M \times M$ invertible matrix. Accordingly, we have the optimal strategy (3.11). Given the fact that $A^\top A$ is the positive semidefinite matrix for any matrix A, the time-dependent component

$$\Lambda^2(t) = \boldsymbol{\mu}_{\boldsymbol{F}}(t)^\top(\boldsymbol{\Sigma}_{\boldsymbol{F}}(t)\boldsymbol{\Sigma}_{\boldsymbol{F}}(t)^\top)^{-1}\boldsymbol{\mu}_{\boldsymbol{F}}(t)$$

is always non-negative.

Substituting $\boldsymbol{\pi}^*$ back, equation (3.12) becomes

$$-\frac{d}{dt}h(t) + \frac{1}{2}\Lambda^2(t)h(t) = 0.$$

Accordingly, we have

$$h(t) = \exp\left(-\frac{1}{2}\int_t^{\widetilde{T}}\Lambda^2(s)ds\right).$$

□

Then, for a given wealth w, the candidate solution u is a non-increasing function with respect to the time t. The solution to the HJB equation is not sufficient if a verification theorem is not proven. The verification theorem connects the HJB equation (3.8) to the control problem of maximizing the expected utility at the terminal time defined in (3.7). Next, we provide the verification theorem for our problem.

Theorem 3.2. *Let $u(t, w)$ be given by (3.10). Then,*

(1) *$u(t, w) \geq V(t, w)$ for all $t \in [0, \widetilde{T}]$ and $w \in \mathbb{R}$,*
(2) *the maximizer $\boldsymbol{\pi}^*$ given by (3.11) is admissible. Therefore, $u(t, w) = V(t, w)$ for all $t \in [0, \widetilde{T}]$ and $w \in \mathbb{R}$, and $\boldsymbol{\pi}^*$ is an optimal strategy.*

Proof. We need to show that for any admissible $\boldsymbol{\pi}$ such that $\partial_t u + \mathcal{L}^{\boldsymbol{\pi}} u \leq 0$, the expected utility of terminal wealth will be less than or equal to what the value function would indicate, namely,

$$\mathbb{E}^{\mathbb{P}}[U(W_{\tilde{T}}^{\boldsymbol{\pi}})|\mathcal{F}_t] \leq u(t, W_t),$$

and that the equality holds when the wealth process is controlled optimally by $\boldsymbol{\pi}^*$; that is,

$$\begin{aligned} V(t, W_t^{\boldsymbol{\pi}^*}) &= \sup_{\boldsymbol{\pi}\in\mathcal{A}_t} \mathbb{E}^{\mathbb{P}}[U(W_{\tilde{T}}^{\boldsymbol{\pi}})|\mathcal{F}_t] \\ &= \mathbb{E}^{\mathbb{P}}[U(W_{\tilde{T}}^{\boldsymbol{\pi}^*})|\mathcal{F}_t] \\ &= u(t, W_t^{\boldsymbol{\pi}^*}). \end{aligned}$$

Using Ito's formula, we obtain

$$du(t, W_t^{\boldsymbol{\pi}}) = (\partial_t + \mathcal{L}^{\boldsymbol{\pi}})u(t, W_t^{\boldsymbol{\pi}})dt + \partial_w u(t, W_t^{\boldsymbol{\pi}})\boldsymbol{\pi}_t^\top \boldsymbol{\Sigma}_{\boldsymbol{F}}(t)d\boldsymbol{Z}_t^{\mathbb{P}}.$$

Notice that $u < 0$ and $\partial_w u = -\gamma u$. Using these, we obtain the inequality

$$\begin{aligned} &d\log(-u(t, W_t^{\boldsymbol{\pi}})) \\ &= \left(-\frac{\gamma^2}{2}\boldsymbol{\pi}_t^\top \boldsymbol{\Sigma}_{\boldsymbol{F}}(t)\boldsymbol{\Sigma}_{\boldsymbol{F}}(t)^\top \boldsymbol{\pi}_t + \frac{(\partial_t + \mathcal{L}^{\boldsymbol{\pi}})u(t, W_t^{\boldsymbol{\pi}})}{u(t, W_t^{\boldsymbol{\pi}})}\right) dt - \gamma\boldsymbol{\pi}_t^\top \boldsymbol{\Sigma}_{\boldsymbol{F}}(t)d\boldsymbol{Z}_t^{\mathbb{P}} \\ &\geq -\frac{\gamma^2}{2}\boldsymbol{\pi}_t^\top \boldsymbol{\Sigma}_{\boldsymbol{F}}(t)\boldsymbol{\Sigma}_{\boldsymbol{F}}(t)^\top \boldsymbol{\pi}_t dt - \gamma\boldsymbol{\pi}_t^\top \boldsymbol{\Sigma}_{\boldsymbol{F}}(t)d\boldsymbol{Z}_t^{\mathbb{P}}. \end{aligned}$$

The last inequality holds due to the fact that $(\partial_t + \mathcal{L}^{\boldsymbol{\pi}})u \leq 0$.

Then, with the admissible strategy satisfying the Novikov condition (3.6), we obtain

$$\begin{aligned} \mathbb{E}^{\mathbb{P}}\left[U(W_{\tilde{T}}^{\boldsymbol{\pi}})|\mathcal{F}_t\right] &= \mathbb{E}^{\mathbb{P}}\left[u(\tilde{T}, W_{\tilde{T}}^{\boldsymbol{\pi}})|\mathcal{F}_t\right] \\ &\leq u(t, W_t^{\boldsymbol{\pi}})\mathcal{E}\left(\int_t^{\tilde{T}} -\gamma\boldsymbol{\pi}_s^\top \boldsymbol{\Sigma}_{\boldsymbol{F}}(s)d\boldsymbol{Z}_s^{\mathbb{P}}\right) \\ &= u(t, W_t^{\boldsymbol{\pi}}), \end{aligned}$$

where $\mathcal{E}(\cdot)$ denotes Doléans–Dade exponential. The equality holds if

$$\partial_t u + \mathcal{L}^{\pi} u = 0.$$

Finally, since $\boldsymbol{\pi}^*$ is a time-deterministic function, it is also admissible. Therefore, $u(t, w)$ is indeed the solution of our control problem. □

We insert the optimal strategy (3.11) into the wealth process (3.5) to derive the SDE for the optimal wealth process

$$dW_t^* = \frac{1}{\gamma}\Lambda^2(t)dt + \frac{1}{\gamma}\boldsymbol{\mu}_{\boldsymbol{F}}(t)^\top(\boldsymbol{\Sigma}_{\boldsymbol{F}}(t)\boldsymbol{\Sigma}_{\boldsymbol{F}}(t)^\top)^{-1}\boldsymbol{\Sigma}_{\boldsymbol{F}}(t)d\boldsymbol{Z}_t^{\mathbb{P}}.$$

From this, we see that the drift of the optimal wealth process is always positive. It is proportional to $\Lambda^2(t)$ defined in (3.9). Both the drift and volatility terms are inversely proportional to risk aversion γ.

In order to quantify the value of trading futures to the investor, we define the investor's certainty equivalent associated with the utility maximization problem. The certainty equivalent is the guaranteed cash amount that would yield the same utility as that from dynamically trading futures according to (3.7). This amounts to applying the inverse of the utility function to the value function in (3.10). Precisely, we define

$$CE(t, w) = w + \frac{1}{2\gamma}\int_t^{\widetilde{T}} \Lambda^2(s)ds. \tag{3.13}$$

Therefore, the certainty equivalent is the sum of the investor's wealth w and a non-negative time-dependent component $\frac{1}{2\gamma}\int_t^{\widetilde{T}} \Lambda^2(s)ds$. The certainty equivalent is also inversely proportional to the risk aversion parameter γ. This means that, all else being equal, a more risk averse investor values the futures trading opportunity less. From (2.9), (3.1)–(3.4), and (3.9), we know the certainty equivalent depends on the coefficient matrix $\boldsymbol{K}$, volatility matrix $\boldsymbol{\Sigma}$, correlation matrix $\boldsymbol{C}$ and market prices of risk $\boldsymbol{\lambda}$. Nevertheless, the certainty equivalent does not depend on the current values of the spot price and factors represented by $\boldsymbol{X}_t$.

3.3 Two-Factor Models

In this section, we discuss the application of our framework to two well-known two-factor models: the Schwartz model and CTOU model. In both cases, Theorem 3.1 can be applied directly, and we state the optimal strategies explicitly using (3.11).

3.3.1 *Schwartz model*

To specify the Schwartz model in our multifactor framework, we set the coefficients in SDE (2.16) and (2.17) to be

$$\boldsymbol{\mu} = \begin{bmatrix} \mu_1 - \sigma_1^2/2 \\ \kappa\alpha \end{bmatrix}, \quad \boldsymbol{K} = \begin{bmatrix} 0 & 1 \\ 0 & \kappa \end{bmatrix},$$

$$\boldsymbol{\Sigma} = \begin{bmatrix} \sigma_1 & 0 \\ 0 & \sigma_2 \end{bmatrix}, \quad \boldsymbol{C} = \begin{bmatrix} 1 & 0 \\ \rho & \sqrt{1-\rho^2} \end{bmatrix},$$

and

$$\boldsymbol{\lambda} = (\mu_1 - r, \lambda_2)^\top.$$

Recall that μ_1 and α are the long-run means, κ is the speed of mean reversion, and λ_2 is the market price of convenience yield risk. The instantaneous correlation between the commodity price and its convenience yield is $\rho \in (-1, 1)$.

Next, we consider the trading problem that investor can choose to trade one futures contract or two different futures contracts. In our framework, we directly apply Theorem 3.1. For the portfolio with T_1-futures only, we apply (3.1), (3.2), (3.3), and (3.4) to get the coefficients for the futures price SDE:

$$\boldsymbol{\mu_F}(t) = \mu_1 - r + \frac{e^{-(T_1-t)\kappa} - 1}{\kappa}\lambda_2,$$

and

$$\boldsymbol{\Sigma_F}(t) = \left(\sigma_1 + \rho\sigma_2 \frac{e^{-(T_1-t)\kappa} - 1}{\kappa}, \quad \sigma_2\sqrt{1-\rho^2}\frac{e^{-(T_1-t)\kappa} - 1}{\kappa}\right).$$

Applying these to formula (3.11), the optimal strategy is given by

$$\begin{aligned} \pi^{(1)*}(t) &= \frac{1}{\gamma} \frac{\boldsymbol{\mu_F}(t)}{\boldsymbol{\Sigma_F}(t)\boldsymbol{\Sigma_F^\top}(t)} \\ &= \frac{1}{\gamma} \frac{\kappa^2(\mu_1 - r) + \kappa(e^{-(T_1-t)\kappa} - 1)\lambda_2}{\kappa^2\sigma_1^2 + 2\rho\sigma_1\sigma_2\kappa(e^{-(T_1-t)\kappa} - 1) + \sigma_2^2\left(e^{-(T_1-t)\kappa} - 1\right)^2}. \end{aligned}$$

As for the portfolio of both two futures, T_1-futures and T_2-futures, according to (3.3), we have

$$\boldsymbol{\mu_F}(t) = \left(\mu_1 - r + \frac{e^{-(T_1-t)\kappa} - 1}{\kappa}\lambda_2, \quad \mu_1 - r + \frac{e^{-(T_2-t)\kappa} - 1}{\kappa}\lambda_2\right)^\top,$$

and

$$\boldsymbol{\Sigma_F}(t) = \begin{bmatrix} \sigma_1 + \rho\sigma_2 \frac{e^{-(T_1-t)\kappa}-1}{\kappa} & \sigma_2\sqrt{1-\rho^2}\frac{e^{-(T_1-t)\kappa}-1}{\kappa} \\ \sigma_1 + \rho\sigma_2 \frac{e^{-(T_2-t)\kappa}-1}{\kappa} & \sigma_2\sqrt{1-\rho^2}\frac{e^{-(T_2-t)\kappa}-1}{\kappa} \end{bmatrix}.$$

Then, applying (3.11) yields the optimal strategies

$$\begin{aligned}\pi^{(1)*}(t) = &- e^{\kappa(T_1-t)}\Big(\left(e^{\kappa t} - e^{\kappa T_2}\right)(r-\mu_1)\sigma_2^2 \\ &+ \left(e^{\kappa t}\lambda_2 + e^{\kappa T_2}(r\kappa - \lambda_2 - \kappa\mu_1)\right)\rho\sigma_1\sigma_2 + e^{\kappa T_2}\kappa\lambda_2\sigma_1^2\Big)/ \\ &\left(\gamma\left(e^{\kappa T_1} - e^{\kappa T_2}\right)(1-\rho^2)\sigma_1^2\sigma_2^2\right),\end{aligned}$$

$$\begin{aligned}\pi^{(2)*}(t) = &\, e^{\kappa(T_2-t)}\Big(\left(e^{\kappa t} - e^{\kappa T_1}\right)(r-\mu_1)\sigma_2^2 \\ &+ \left(e^{\kappa t}\lambda_2 + e^{\kappa T_1}(r\kappa - \lambda_2 - \kappa\mu_1)\right)\rho\sigma_1\sigma_2 + e^{\kappa T_1}\kappa\lambda_2\sigma_1^2\Big)/ \\ &\left(\gamma\left(e^{\kappa T_1} - e^{\kappa T_2}\right)(1-\rho^2)\sigma_1^2\sigma_2^2\right).\end{aligned}$$

The results above are also obtained by Leung and Yan (2019) using a different method.

3.3.2 *CTOU model*

This model is called the CTOU, studied by Mencia and Sentana (2013) for pricing VIX futures. Later, Leung and Yan (2018) analyze the futures portfolio optimization problem under this model. The CTOU process also belongs to our multifactor framework. Indeed, this amounts to setting the coefficients in SDE (2.3) and (2.5):

$$\boldsymbol{\mu} = \begin{bmatrix} \lambda_1 \\ \kappa_2\theta + \lambda_2 \end{bmatrix}, \quad \boldsymbol{K} = \begin{bmatrix} \kappa_1 & -\kappa_1 \\ 0 & \kappa_2 \end{bmatrix}, \tag{3.14}$$

$$\boldsymbol{\Sigma} = \begin{bmatrix} \sigma_1 & 0 \\ 0 & \sigma_2 \end{bmatrix}, \quad \boldsymbol{C} = \begin{bmatrix} 1 & 0 \\ \rho & \sqrt{1-\rho^2} \end{bmatrix},$$

and

$$\boldsymbol{\lambda} = (\lambda_1, \lambda_2)^\top, \tag{3.15}$$

where the mean θ, speeds of mean reversion $\{\kappa_1, \kappa_2\}$, the volatility parameters $\{\sigma_1, \sigma_2\}$, and adjusted market prices of risk $\{\lambda_1, \lambda_2\}$ are all constants.

In Mencia and Sentana (2013) and Leung and Yan (2018), the instantaneous correlation ρ is set to be 0.

Then, consider two futures contracts $F^{(1)}$ and $F^{(2)}$, written on this underlying, whose maturity are T_1 and T_2, respectively. Recall that Proposition 2.2 gives the futures price formula:

$$F^{(k)}(t, \boldsymbol{x}) = \exp\left(\boldsymbol{a}^{(k)}(t)^\top \boldsymbol{x} + \beta^{(k)}(t)\right),$$

where $\boldsymbol{a}^{(k)}(t)$ satisfies

$$\boldsymbol{a}^{(k)}(t) = \left(e^{-(T_k-t)\kappa_1}, \frac{\kappa_1}{\kappa_1 - \kappa_2}\left(e^{-(T_k-t)\kappa_2} - e^{-(T_k-t)\kappa_1}\right)\right)^\top,$$

and $\beta^{(k)}(t)$ follows:

$$\begin{aligned}\beta^{(k)}(t) &= \theta - D_1(T_k - t)\theta + \frac{\sigma_1^2}{4\kappa_1}\left(1 - e^{-2\kappa_1(T_k-t)}\right)\\&\quad + \rho\sigma_1\sigma_2\frac{\kappa_1}{\kappa_1-\kappa_2}\left(\frac{1-e^{-(\kappa_1+\kappa_2)(T_k-t)}}{\kappa_1+\kappa_2} - \frac{1-e^{-2\kappa_1(T_k-t)}}{2\kappa_1}\right)\\&\quad + \frac{\sigma_2^2}{2}\left(\frac{\kappa_1}{\kappa_1-\kappa_2}\right)^2\left(\frac{1-e^{-2\kappa_2(T_k-t)}}{2\kappa_2} + \frac{1-e^{-2\kappa_1(T_k-t)}}{2\kappa_1}\right.\\&\quad \left. - 2\frac{1-e^{-(\kappa_1+\kappa_2)(T_k-t)}}{\kappa_1+\kappa_2}\right),\end{aligned}$$

with

$$D_1(\tau) = \frac{\kappa_1}{\kappa_1 - \kappa_2}e^{-\kappa_2\tau} - \frac{\kappa_2}{\kappa_1 - \kappa_2}e^{-\kappa_1\tau}.$$

Next, we consider the trading problem for the investor. The investor can choose to trade one futures contract or two different futures contracts. This case is covered by Theorem 3.1, which can be applied directly.

If only futures contract $F^{(1)}$ is included in the portfolio, then according to (3.1) and (3.3), we have $\boldsymbol{\mu_F}(t) \equiv \mu_F(t)$, where

$$\mu_F(t) = e^{-(T_1-t)\kappa_1}\lambda_1 + D_2(T_1 - t)\lambda_2$$

is a scalar, and

$$\boldsymbol{\Sigma_F}(t) = \left(e^{-(T_1-t)\kappa_1}\sigma_1 + \rho D_2(T_1 - t)\sigma_2, \quad \sqrt{1-\rho^2}D_2(T_1 - t)\sigma_2\right),$$

with

$$D_2(\tau) = \frac{\kappa_1}{\kappa_1 - \kappa_2}\left(e^{-\tau\kappa_2} - e^{-\tau\kappa_1}\right).$$

Then, the optimal strategy (3.11) becomes

$$\begin{aligned}\pi^{(1)*}(t) &= \frac{1}{\gamma}\frac{\mu_F(t)}{\boldsymbol{\Sigma}_{\boldsymbol{F}}(t)\boldsymbol{\Sigma}_{\boldsymbol{F}}^\top(t)}\\ &= \frac{1}{\gamma}\frac{e^{-(T_1-t)\kappa_1}\lambda_1 + D_2(T_1-t)\lambda_2}{e^{-2(T_1-t)\kappa_1}\sigma_1^2 + D_2^2(T_1-t)\sigma_2^2 + 2\rho e^{-(T_1-t)\kappa_1}D_2(T_1-t)\sigma_1\sigma_2}.\end{aligned} \tag{3.16}$$

By Theorem 3.1, we can also get

$$\begin{aligned}\Lambda_1^2(t) &= \gamma\pi^{(1)*}(t)\mu_F(t)\\ &= \frac{\left(e^{-(T_1-t)\kappa_1}\lambda_1 + D_2(T_1-t)\lambda_2\right)^2}{e^{-2(T_1-t)\kappa_1}\sigma_1^2 + D_2^2(T_1-t)\sigma_2^2 + 2\rho e^{-(T_1-t)\kappa_1}D_2(T_1-t)\sigma_1\sigma_2},\end{aligned} \tag{3.17}$$

and certainty equivalent

$$CE_1(t,w) = w + \frac{1}{2\gamma}\int_t^{\tilde{T}} \Lambda_1^2(s)ds. \tag{3.18}$$

As for trading two futures contracts $F^{(1)}$ and $F^{(2)}$, according to (3.3), we have

$$\boldsymbol{\mu}_{\boldsymbol{F}}(t) = \begin{bmatrix} e^{-(T_1-t)\kappa_1}\lambda_1 + D_2(T_1-t)\lambda_2 \\ e^{-(T_2-t)\kappa_1}\lambda_1 + D_2(T_2-t)\lambda_2 \end{bmatrix}, \tag{3.19}$$

and

$$\boldsymbol{\Sigma}_{\boldsymbol{F}}(t) = \begin{bmatrix} e^{-(T_1-t)\kappa_1}\sigma_1 + \rho D_2(T_1-t)\sigma_2 & \sqrt{1-\rho^2}D_2(T_1-t)\sigma_2 \\ e^{-(T_2-t)\kappa_1}\sigma_1 + \rho D_2(T_2-t)\sigma_2 & \sqrt{1-\rho^2}D_2(T_2-t)\sigma_2 \end{bmatrix}. \tag{3.20}$$

Then, applying formula (3.11), we obtain the optimal strategies

$$\pi_1^*(t) = \big(\kappa_1\big(e^{-(T_2-t)\kappa_2} - e^{-(T_2-t)\kappa_1}\big)(\sigma_2^2\lambda_1 - \rho\sigma_1\sigma_2\lambda_2) \\ + (\kappa_1-\kappa_2)e^{-(T_2-t)\kappa_1}(\rho\sigma_1\sigma_2\lambda_1 - \sigma_1^2\lambda_2)\big)/ \\ \big(\gamma\kappa_1(1-\rho^2)\big(e^{-(T_1-t)\kappa_1-(T_2-t)\kappa_2} - e^{-(T_1-t)\kappa_2-(T_2-t)\kappa_1}\big)\sigma_1^2\sigma_2^2\big), \tag{3.21}$$

$$\pi_2^*(t) = -\big(\kappa_1\big(e^{-(T_1-t)\kappa_2} - e^{-(T_1-t)\kappa_1}\big)(\sigma_2^2\lambda_1 - \rho\sigma_1\sigma_2\lambda_2) \\ + (\kappa_1-\kappa_2)e^{-(T_1-t)\kappa_1}(\rho\sigma_1\sigma_2\lambda_1 - \sigma_1^2\lambda_2)\big)/ \\ \big(\gamma\kappa_1(1-\rho^2)\big(e^{-(T_1-t)\kappa_1-(T_2-t)\kappa_2} - e^{-(T_1-t)\kappa_2-(T_2-t)\kappa_1}\big)\sigma_1^2\sigma_2^2\big). \tag{3.22}$$

As a check, when we set $\rho = 0$ in (3.16), (3.21) and (3.22), we recover the optimal strategies provided in Leung and Yan (2018).

We note that by substituting (3.19) and (3.20) in (3.9) and (3.10), we obtain

$$\Lambda_2^2(t) = \frac{\lambda_2^2\sigma_1^2 - 2\rho\lambda_1\lambda_2\sigma_1\sigma_2 + \lambda_1^2\sigma_2^2}{(1-\rho^2)\sigma_1^2\sigma_2^2} \tag{3.23}$$

and the value function:

$$u_2(t,w) = -\exp\left(-\gamma w - \frac{1}{2}\frac{\lambda_2^2\sigma_1^2 - 2\rho\lambda_1\lambda_2\sigma_1\sigma_2 + \lambda_1^2\sigma_2^2}{(1-\rho^2)\sigma_1^2\sigma_2^2}(\widetilde{T}-t)\right),$$

which is surprisingly independent of the speeds of mean reversion κ_1 and κ_2. Furthermore, if we represent the value function using the risk premia $\boldsymbol{\zeta} = (\zeta_1, \zeta_2)^\top$, where $\boldsymbol{\zeta}$ satisfies

$$\boldsymbol{\zeta} = \boldsymbol{C}^{-1}\boldsymbol{\Sigma}^{-1}\boldsymbol{\lambda},$$

then the value function is simplified as

$$u_2(t,w) = -\exp\left(-\gamma w - \frac{\zeta_1^2 + \zeta_2^2}{2}(\widetilde{T}-t)\right), \tag{3.24}$$

which reveals that the value function depends only on the risk premia. This is an interesting contrast when compared to the case of trading only a single futures, where the certainty equivalent, $CE_1(t,w)$ in (3.18), does not admit the same kind of simplification and depends on other model parameters. This is intuitive since trading one futures exposes the investor to unhedged risk in the model, so it is reasonable that the trading strategy should depend on the model parameters.

In turn, the certainty equivalent is given by

$$CE_2(t, w) = w + \frac{1}{2\gamma}\frac{\lambda_2^2\sigma_1^2 - 2\rho\lambda_1\lambda_2\sigma_1\sigma_2 + \lambda_1^2\sigma_2^2}{(1-\rho^2)\sigma_1^2\sigma_2^2}(\widetilde{T} - t)$$
$$= w + \frac{\zeta_1^2 + \zeta_2^2}{2\gamma}(\widetilde{T} - t). \tag{3.25}$$

Therefore, the value of trading opportunity is proportional to the squared sum of the two risk premia $(\zeta_1^2 + \zeta_2^2)$ associated with the two Brownian motions in the CTOU model. In particular, this means that if the price dynamics of the spot asset under the historical measure and risk-neutral measure are identical, which corresponds to the case of zero risk premia, then the certainty equivalent will vanish. Lastly, as expected, it is also inversely proportional to risk aversion γ.

In the following proposition, we show that it is more beneficial for the investor to trade two futures instead of one.

Proposition 3.3. *For any time t and wealth w, the following inequality holds:*

$$CE_2(t, w) \geq CE_1(t, w),$$

where $CE_1(t, w)$ and $CE_2(t, w)$ are given by (3.18) *and* (3.25), *respectively.*

Proof. Both certainty equivalents admit the same form given by (3.13), so it remains to prove that $\Lambda_1^2(t) \leq \Lambda_2^2(t)$ for any time $t \leq \tilde{T}$. Denote

$$a = \exp(-(T_1 - t)\kappa_1)$$
$$b = D_2(T_1 - t).$$

Then by (3.17), we have

$$\Lambda_1^2(t) = \frac{(a\lambda_1 + b\lambda_2)^2}{a^2\sigma_1^2 + b^2\sigma_2^2 + 2\rho ab\sigma_1\sigma_2}.$$

Using this and (3.23), we have

$$\Lambda_2^2(t) - \Lambda_1^2(t) = \frac{\left(\sigma_1^2\lambda_2 a - \sigma_2^2\lambda_1 b + \rho\sigma_1\sigma_2(b\lambda_2 - a\lambda_1)\right)^2}{(1-\rho^2)(a^2\sigma_1^2 + b^2\sigma_2^2 + 2\rho ab\sigma_1\sigma_2)\sigma_1^2\sigma_2^2} \geq 0,$$

as desired. □

Numerical examples of the certainty equivalents for CTOU model are presented in Section 3.5.

3.4 The Multiscale Central Tendency Ornstein–Uhlenbeck Model

In this section, we introduce the class of two-scale CTOU process. We also discuss the futures pricing and futures trading problem under this model.

3.4.1 *Model formulation*

Following the change of measure procedure described in Section 2.1, we denote the log-price of the underlying asset by $X_t^{(1)}$ and its evolution under the physical measure $\mathbb{P}$ is given by the system of stochastic differential equations:

$$\begin{aligned} dX_t^{(1)} &= \kappa\big(X_t^{(2)} + X_t^{(3)} - X_t^{(1)}\big)dt + \sigma_1 dZ_t^{\mathbb{P},1}, \\ dX_t^{(2)} &= \frac{1}{\epsilon}\big(\alpha_2 - X_t^{(2)}\big)dt + \frac{1}{\sqrt{\epsilon}}\sigma_2\left(\rho_{12}dZ_t^{\mathbb{P},1} + \sqrt{1-\rho_{12}^2}\,dZ_t^{\mathbb{P},2}\right), \\ dX_t^{(3)} &= \delta\big(\alpha_3 - X_t^{(3)}\big)dt \\ &\quad + \sqrt{\delta}\sigma_3\left(\rho_{13}dZ_t^{\mathbb{P},1} + \rho_{23}dZ_t^{\mathbb{P},2} + \sqrt{1-\rho_{13}^2-\rho_{23}^2}\,dZ_t^{\mathbb{P},3}\right), \end{aligned} \tag{3.26}$$

where $Z^{\mathbb{P},1}$, $Z^{\mathbb{P},2}$, and $Z^{\mathbb{P},3}$ are independent Brownian motions under the physical measure $\mathbb{P}$. The correlation coefficient ρ_{12} is constant with $|\rho_{12}| < 1$.

Recall that the stochastic mean of log price $X^{(1)}$ is the sum of the fast mean-reverting factor $X_t^{(2)}$ and the slow varying factor $X_t^{(3)}$. The correlation coefficient ρ_{13} and ρ_{23} are constants such that $\rho_{13}^2 + \rho_{23}^2 < 1$.

We specify the risk premium as ζ_i, for $i = 1, 2, 3$, which satisfy

$$dZ_t^{\mathbb{Q},i} = dZ_t^{\mathbb{P},i} + \zeta_i,$$

where $Z^{\mathbb{Q},1}$, $Z^{\mathbb{Q},2}$, and $Z^{\mathbb{Q},3}$ are independent Brownian motions under risk-neutral measure $\mathbb{Q}$. We introduce the combined market prices of volatility risk λ_i, for $i = 1, 2, 3$, defined by

$$\begin{aligned} \lambda_1 &= \zeta_1\sigma_1, \\ \lambda_2 &= \frac{1}{\sqrt{\epsilon}}\sigma_2\left(\zeta_1\rho_{12} + \zeta_2\sqrt{1-\rho_{12}^2}\right), \\ \lambda_3 &= \sqrt{\delta}\sigma_3\left(\zeta_1\rho_{13} + \zeta_2\rho_{23} + \zeta_3\sqrt{1-\rho_{13}^2-\rho_{23}^2}\right). \end{aligned}$$

Then, we write the evolution under the risk neutral measure as

$$
\begin{aligned}
dX_t^{(1)} &= \kappa\big(X_t^{(2)} + X_t^{(3)} - X_t^{(1)} - \lambda_1/\kappa\big)dt + \sigma_1 dZ_t^{\mathbb{Q},1},\\
dX_t^{(2)} &= \frac{1}{\epsilon}\big(\alpha_2 - X_t^{(2)} - \epsilon\lambda_2\big)dt + \frac{1}{\sqrt{\epsilon}}\sigma_2\left(\rho_{12}dZ_t^{\mathbb{Q},1} + \sqrt{1-\rho_{12}^2}dZ_t^{\mathbb{Q},2}\right),\\
dX_t^{(3)} &= \delta\big(\alpha_3 - X_t^{(3)} - \lambda_3/\delta\big)dt\\
&\qquad + \sqrt{\delta}\sigma_3\left(\rho_{13}dZ_t^{\mathbb{Q},1} + \rho_{23}dZ_t^{\mathbb{Q},2} + \sqrt{1-\rho_{13}^2-\rho_{23}^2}dZ_t^{\mathbb{Q},3}\right).
\end{aligned}
$$

This two-scale CTOU model belongs to our multifactor model framework. Indeed, this amounts to setting the coefficients in SDE (2.3) and (2.5) to be

$$
\boldsymbol{\mu} = (0, \alpha_2/\epsilon, \delta\alpha_3)^\top, \quad \boldsymbol{\lambda} = (\lambda_1, \lambda_2, \lambda_3)^\top,
$$

$$
\boldsymbol{K} = \begin{bmatrix} \kappa & -\kappa & -\kappa \\ 0 & 1/\epsilon & 0 \\ 0 & 0 & \delta \end{bmatrix}, \quad \boldsymbol{\Sigma} = \begin{bmatrix} \sigma_1 & 0 & 0 \\ 0 & \sigma_2/\sqrt{\epsilon} & 0 \\ 0 & 0 & \sqrt{\delta}\sigma_3 \end{bmatrix},
$$

and

$$
\boldsymbol{C} = \begin{bmatrix} 1 & 0 & 0 \\ \rho_{12} & \sqrt{1-\rho_{12}^2} & 0 \\ \rho_{13} & \rho_{23} & \sqrt{1-\rho_{13}^2-\rho_{23}^2} \end{bmatrix}.
$$

3.4.2 *Futures pricing and futures trading*

Next, we analyze the trading problem for the investor. According to (2.21), which follows from Proposition 2.2, the price function for the futures contract with maturity T_k under this multiscale CTOU model is given by

$$
F^{(k)}(t, \boldsymbol{x}) = \exp\big(\boldsymbol{a}^{(k)}(t)^\top \boldsymbol{x} + \beta^{(k)}(t)\big).
$$

We refer to (2.21) for the expressions of $\boldsymbol{a}^{(k)}(t)$ and $\beta^{(k)}(t)$.

Then, we apply (2.17), (3.1), (3.2) to obtain the $\mathbb{P}$-dynamics of $F_t^{(k)}$:

$$
\frac{dF_t^{(k)}}{F_t^{(k)}} = \mu_F^{(k)}(t)dt + \boldsymbol{\sigma}_F^{(k)}(t)^\top d\boldsymbol{Z}_t^{\mathbb{P}},
$$

where

$$\mu_F^{(k)}(t) = f_k(\kappa,t)\lambda_1 + \frac{\epsilon\kappa}{1-\epsilon\kappa}\left(f_k(\kappa,t) - f_k(1/\epsilon,t)\right)\lambda_2 + \frac{\kappa}{\delta-\kappa}\left(f_k(\kappa,t) - f_k(\delta,t)\right)\lambda_3, \tag{3.27}$$

and

$$\boldsymbol{\sigma}_F^{(k)}(t) = \begin{bmatrix} f_k(\kappa,t)\sigma_1 + \rho_{12}\frac{\epsilon\kappa}{1-\epsilon\kappa}\left(f_k(\kappa,t) - f_k(1/\epsilon,t)\right)\sigma_2 \\ +\rho_{13}\frac{\kappa}{\delta-\kappa}\left(f_k(\kappa,t) - f_k(\delta,t)\right)\sigma_3 \\ \sqrt{1-\rho_{12}^2}\frac{\epsilon\kappa}{1-\epsilon\kappa}\left(f_k(\kappa,t) - f_k(1/\epsilon,t)\right)\sigma_2 \\ +\rho_{23}\frac{\kappa}{\delta-\kappa}\left(f_k(\kappa,t) - f_k(\delta,t)\right)\sigma_3 \\ \sqrt{1-\rho_{13}^2-\rho_{23}^2}\frac{\kappa}{\delta-\kappa}\left(f_k(\kappa,t) - f_k(\delta,t)\right)\sigma_3 \end{bmatrix}. \tag{3.28}$$

As discussed in Section 3.2, there are three options for the investor: trade one futures contract, trade two different futures contracts and trade three different futures contracts.

If there is only one futures contract $F^{(1)}$ available in the market, then applying (3.3), we get

$$\boldsymbol{\mu_F}(t) = \mu_F^{(1)}(t),$$
$$\boldsymbol{\Sigma_F} = \boldsymbol{\sigma}_F^{(1)}(t)^\top.$$

With this, the optimal strategy for trading a single futures contract, according to (3.11), becomes

$$\pi^{(1)*}(t) = \frac{1}{\gamma}\frac{\boldsymbol{\mu_F}(t)}{\boldsymbol{\Sigma_F}(t)\boldsymbol{\Sigma}_{\boldsymbol{F}}^\top(t)} = \frac{1}{\gamma}\frac{\mu_F^{(1)}(t)}{\boldsymbol{\sigma}_F^{(1)}(t)^\top\boldsymbol{\sigma}_F^{(1)}(t)}, \tag{3.29}$$

where $\mu_F^{(1)}(t)$ and $\boldsymbol{\sigma}_F^{(1)}(t)$ are given by (3.27) and (3.28), respectively.

If the investor trade two futures $F^{(1)}$ and $F^{(2)}$, then by (3.3), we have

$$\boldsymbol{\mu_F}(t) = \left(\mu_F^{(1)}(t), \mu_F^{(2)}(t)\right)^\top,$$
$$\boldsymbol{\Sigma_F}(t) = \left(\boldsymbol{\sigma}_F^{(1)}(t), \boldsymbol{\sigma}_F^{(2)}(t)\right)^\top.$$

Then, according to (3.11), we have the optimal strategy

$$\pi^{(1)*}(t) = \frac{\boldsymbol{\sigma}_F^{(2)}(t)^\top \boldsymbol{\sigma}_F^{(2)}(t)\mu_F^{(1)}(t) - \boldsymbol{\sigma}_F^{(1)}(t)^\top \boldsymbol{\sigma}_F^{(2)}(t)\mu_F^{(2)}(t)}{\gamma\left(\boldsymbol{\sigma}_F^{(1)}(t)^\top \boldsymbol{\sigma}_F^{(1)}(t)\boldsymbol{\sigma}_F^{(2)}(t)^\top \boldsymbol{\sigma}_F^{(2)}(t) - \left(\boldsymbol{\sigma}_F^{(1)}(t)^\top \boldsymbol{\sigma}_F^{(2)}(t)\right)^2\right)},$$

$$\pi^{(2)*}(t) = \frac{\boldsymbol{\sigma}_F^{(1)}(t)^\top \boldsymbol{\sigma}_F^{(1)}(t)\mu_F^{(2)}(t) - \boldsymbol{\sigma}_F^{(1)}(t)^\top \boldsymbol{\sigma}_F^{(2)}(t)\mu_F^{(1)}(t)}{\gamma\left(\boldsymbol{\sigma}_F^{(1)}(t)^\top \boldsymbol{\sigma}_F^{(1)}(t)\boldsymbol{\sigma}_F^{(2)}(t)^\top \boldsymbol{\sigma}_F^{(2)}(t) - \left(\boldsymbol{\sigma}_F^{(1)}(t)^\top \boldsymbol{\sigma}_F^{(2)}(t)\right)^2\right)}.$$

If the investor trades three futures $F^{(1)}$, $F^{(2)}$, and $F^{(3)}$, then by (3.3), we have

$$\boldsymbol{\mu}_{\boldsymbol{F}}(t) = \left(\mu_F^{(1)}(t), \mu_F^{(2)}(t), \mu_F^{(3)}(t)\right)^\top,$$

$$\boldsymbol{\Sigma}_{\boldsymbol{F}}(t) = \left(\boldsymbol{\sigma}_F^{(1)}(t), \boldsymbol{\sigma}_F^{(2)}(t), \boldsymbol{\sigma}_F^{(3)}(t)\right)^\top.$$

Then, according to (3.11), we have the optimal strategy

$$\begin{bmatrix} \pi^{(1)*}(t) \\ \pi^{(2)*}(t) \\ \pi^{(3)*}(t) \end{bmatrix} = \frac{1}{\gamma}\begin{bmatrix} \boldsymbol{\sigma}_F^{(1)}(t)^\top \boldsymbol{\sigma}_F^{(1)}(t), & \boldsymbol{\sigma}_F^{(1)}(t)^\top \boldsymbol{\sigma}_F^{(2)}(t), & \boldsymbol{\sigma}_F^{(1)}(t)^\top \boldsymbol{\sigma}_F^{(3)}(t) \\ \boldsymbol{\sigma}_F^{(2)}(t)^\top \boldsymbol{\sigma}_F^{(1)}(t), & \boldsymbol{\sigma}_F^{(2)}(t)^\top \boldsymbol{\sigma}_F^{(2)}(t), & \boldsymbol{\sigma}_F^{(2)}(t)^\top \boldsymbol{\sigma}_F^{(3)}(t) \\ \boldsymbol{\sigma}_F^{(3)}(t)^\top \boldsymbol{\sigma}_F^{(1)}(t), & \boldsymbol{\sigma}_F^{(3)}(t)^\top \boldsymbol{\sigma}_F^{(2)}(t), & \boldsymbol{\sigma}_F^{(3)}(t)^\top \boldsymbol{\sigma}_F^{(3)}(t) \end{bmatrix}^{-1} \boldsymbol{\mu}_{\boldsymbol{F}}(t). \tag{3.30}$$

3.5 Numerical Illustration

We begin with numerical examples for the multiscale CTOU model discussed in Section 3.4. With the closed-form expressions obtained in Section 3.4.2, we now implement the futures prices, optimal strategies and wealth processes numerically, using the parameters in Table 3.1. Primarily, we let ϵ and δ be small parameters and we consider trading three futures with maturities $T_1 = 1/12$ year, $T_2 = 2/12$ year and $T_3 = 3/12$ year. Then, our trading horizon will be $\widetilde{T} = 1/12$ year, no greater than the futures maturities. We use "trading day" as the x axis in some figures. We assume there be 252 trading days in a year and 21 trading days for a month. Therefore, our trading horizon is 21 trading days in total.

Table 3.1. Parameters for the multiscale CTOU model.

$X_0^{(1)}$	$X_0^{(2)}$	$X_0^{(3)}$	α_2	α_3	κ	ϵ	δ	σ_1	σ_2	σ_3
1	0.5	0.5	0.5	0.5	5	0.05	0.01	0.8	0.02	0.3
ρ_{12}	ρ_{13}	ρ_{23}	λ_1	λ_2	λ_3	T_1	T_2	T_3	$\widetilde{T}$	γ
0	0	0	0.02	0.02	0.02	1/12	2/12	3/12	1/12	1

We plot the simulated paths and 95% confidence intervals for three factors in Figures 3.1(a) and 3.1(b). As shown in Figure 3.1(c), the 95% confidence interval of the slow-varying factor $X^{(3)}$ is much narrower than the one for fast-varying factor $X^{(2)}$. At the bottom, we plot the spot price in the solid lines and futures prices in the dashed lines. We observe that three paths for the futures prices are highly correlated and T_1-futures price is equal to the asset's spot price at its maturity date T_1, which is the 21st trading day.

In Figure 3.2, we plot the optimal positions (in dollars) as functions of time for different portfolios. We illustrate the optimal strategies for one-contract portfolio, two-contract portfolio and three-contract portfolio respectively. Solid lines represent the optimal investment on T_1-futures. Dashed lines represent the optimal investment on T_2-futures. Dotted lines represent optimal investment on T_3-futures. The formula for optimal strategies are given in (3.29)–(3.30). As it turns out, the optimal positions in cash amount are deterministic functions for time, but the units of futures in the portfolios still depend directly on asset's spot price and futures prices.

When trading one futures, the investor faces significant risks exposure. Therefore, the position is relatively small due to risk aversion. Moreover, the investor takes significantly larger position in three-contract portfolio since he can hedge all the risks. From a computational perspective, this is related to the large condition number of the matrix in (3.30) where the optimal positions are calculated. Nevertheless, the net exposure is significantly reduced by the opposite (long/short) positions. As shown in Figure 3.3, the net optimal positions across all three portfolios are of the same order of magnitude and very close to zero.

Figure 3.4 shows the wealth processes for three portfolios. We first plot wealth processes as functions of time for one-contract portfolio and two-contract portfolio. We observe that the wealth paths for these two portfolios are pretty close, and they both seem to mostly follow the spot price process,

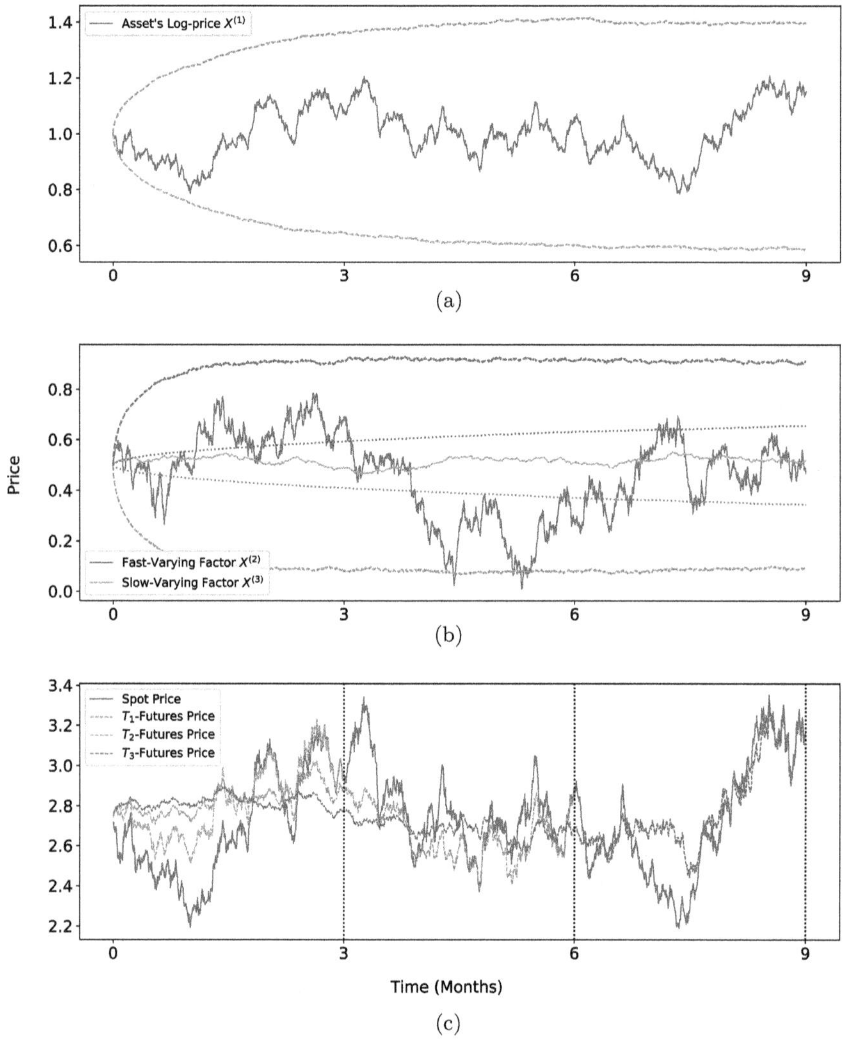

Fig. 3.1. (a) Simulated paths for asset's log-price $X^{(1)}$. Dashed curves represent 95% confidence intervals. (b) Simulated path for the fast-varying factor $X^{(2)}$ and slow-varying factor $X^{(2)}$. Dashed curves represent 95% confidence intervals for $X^{(2)}$ and dotted curves represent 95% confidence intervals for $X^{(3)}$. (c) Simulated paths for asset's spot price and futures prices. Parameters are shown in Table 3.1.

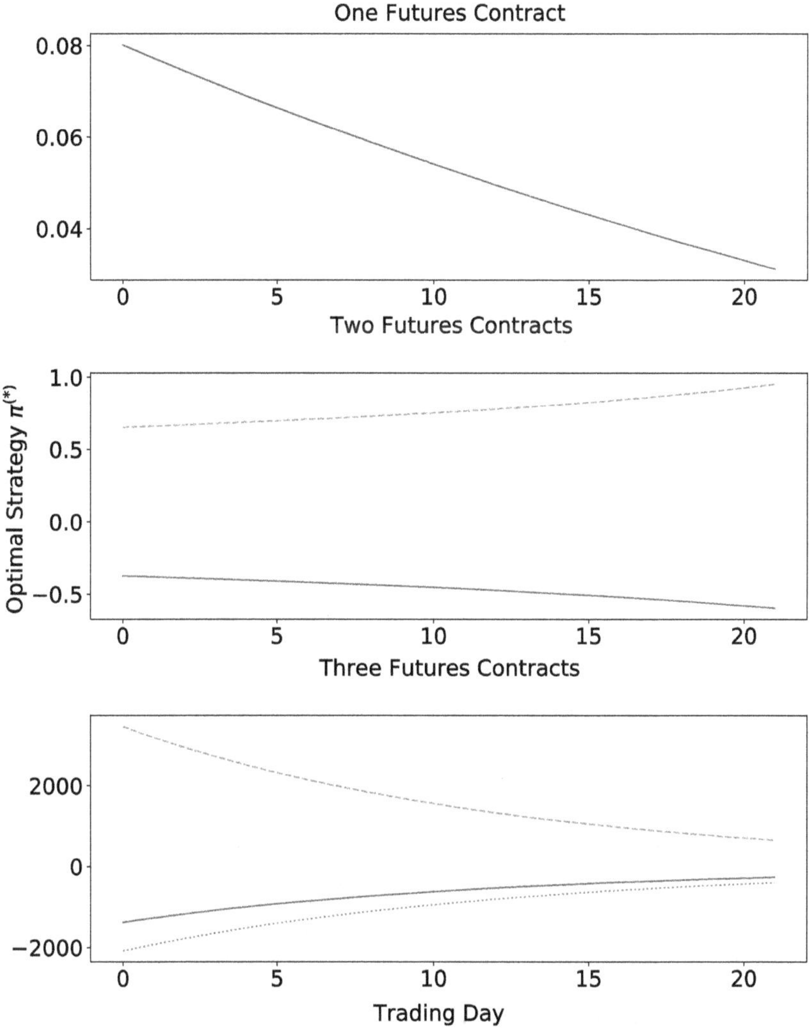

Fig. 3.2. The optimal strategy $\pi^{(*)}$ as a function of time for different futures portfolios. In each plot when applicable, the solid line is the optimal investment on T_1-futures, dashed line is the optimal investment on T_2-futures, and dotted line is the optimal investment on T_3-futures. Parameters are shown in Table 3.1.

which is also trending downward in a similar manner. However, the wealth process for the three-contract portfolio exhibits very different path behavior compared to those on the left. In particular, this wealth process does not seem to follow the spot price.

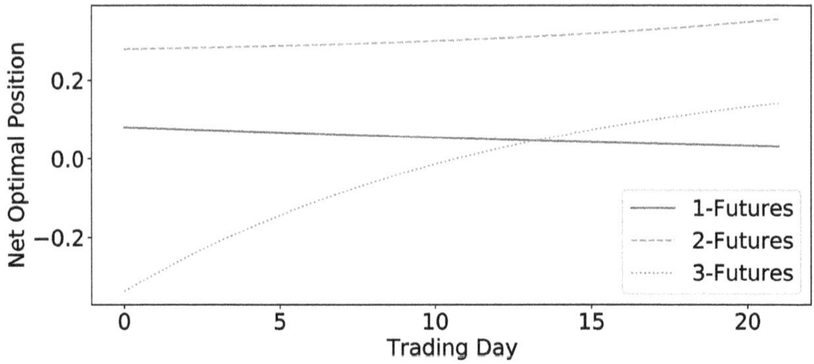

Fig. 3.3. The net optimal positions for the futures portfolios with different numbers of futures. The maximum and minimum net optimal positions for the one-futures portfolio (solid line) are 0.08 and 0.03, respectively. The maximum and minimum net optimal positions for the two-futures portfolio (dashed line) are 0.36 and 0.28, respectively. The maximum and minimum net optimal positions for the three-futures portfolio (dotted line) are 0.14 and −0.34, respectively.

We plot certainty equivalent for different portfolios in Figure 3.5. First, we observe that the certainty equivalent increases with respect to the trading horizon $\widetilde{T}$. As the trading horizon reduces to zero, the certainty equivalent converges to the initial wealth w, which is set to be 0 in this example. It is consistent with equation (3.13). Second, with more contracts to trade, the investor has a higher certainty equivalent, but different futures combinations lead to varying values. In this example, trading the short-term futures, T_1 and T_2 futures, yields the highest certainty equivalent. On the right, we see that risk aversion reduces the certainty equivalent for any trading horizon $\widetilde{T}$.

In Table 3.2, we present the certainty equivalents ($\times 10^{-4}$) for all possible futures combinations and different correlation parameters. Other parameters are shown in Table 3.1. In this table, nine different correlation configurations under the CTOU model are shown. As we can see, the certainty equivalent is much higher when there are more contracts to trade. In addition, if there is only one futures contract to trade, the certainty equivalent is increasing with respect to its maturity, see first three columns. In addition, the certainly equivalents depend on the correlation parameters ρ_{12} and ρ_{13}.

Lastly, we return to the CTOU model presented in Section 3.3.2. In Table 3.3, we present the certainty equivalents corresponding to different values of the speed of reversion κ_1 and correlation ρ. The certainty

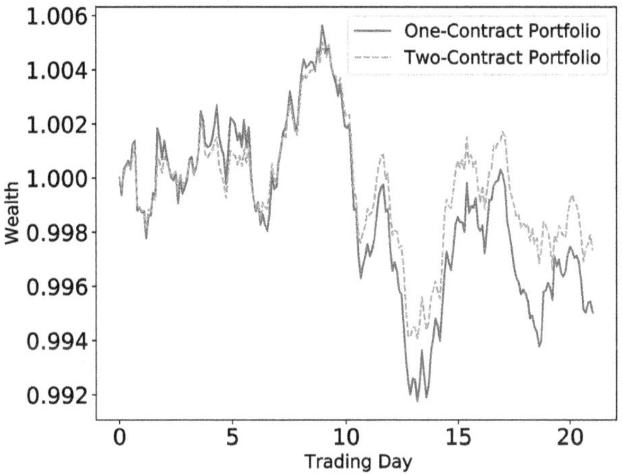

(a)

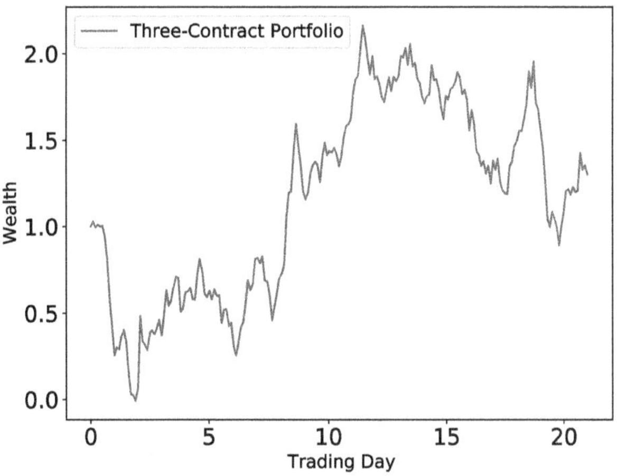

(b)

Fig. 3.4. (a) Wealth processes as functions of time for one-contract portfolio and two-contract portfolio. (b) Wealth process as function of time for three contract-portfolio. Parameters are shown in Table 3.1.

equivalents ($\times 10^{-3}$) for all possible futures combinations, different speed of reversion κ_1 and correlation parameters ρ under the CTOU model. Other parameters are: $\kappa_2 = 0.3$, $\sigma_1 = 1.037$, $\sigma_2 = 0.446$, $\zeta_1 = -0.010$, $\zeta_2 = 2.242$, $T_1 = 1/12$, $T_2 = 2/12$, and $\widetilde{T} = 1/12$.

We use the estimated parameters from the "full sample" in Table 4 of Mencia and Sentana (2013) and Table 1 of Leung and Yan (2018). It shows

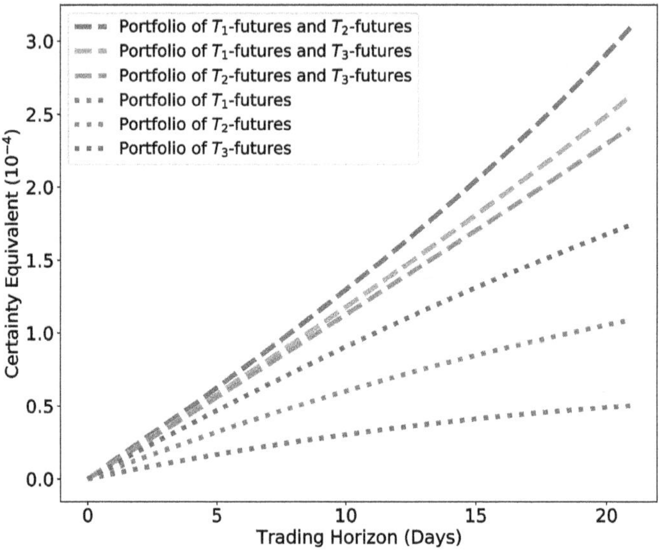

(a)

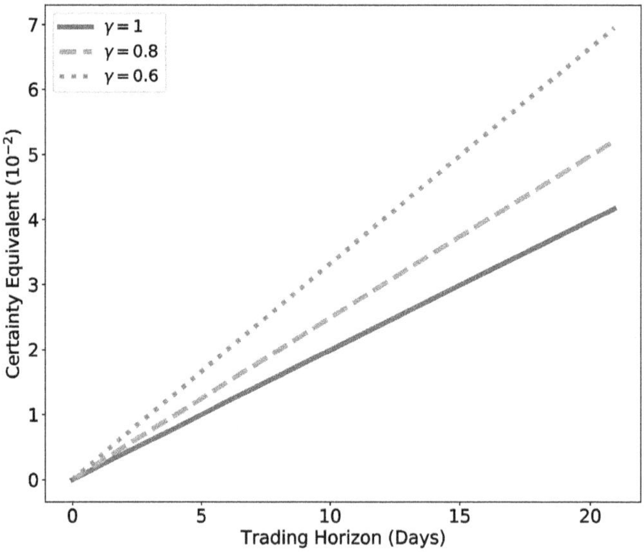

(b)

Fig. 3.5. (a) Certainty equivalents as functions of trading horizon $\widetilde{T}$ for one-contract portfolios and two-contract portfolios. (b) Certainty equivalents as functions of trading horizon $\widetilde{T}$ for three-contract portfolio with different risk-aversion parameter γ. Other parameters are shown in Table 3.1.

Table 3.2. Certainty equivalents of futures portfolios with three available contracts of different maturities.

	Futures combinations (maturity)						
Parameters	T_1	T_2	T_3	$\{T_1, T_2\}$	$\{T_1, T_3\}$	$\{T_2, T_3\}$	$\{T_1, T_2, T_3\}$
$\rho_{12} = 0$							
$\rho_{13} = -0.5$	0.563	1.58	3.25	5.36	4.65	4.41	419
$\rho_{13} = 0$	0.502	1.09	1.74	3.09	2.62	2.41	417
$\rho_{13} = 0.5$	0.456	0.837	1.19	2.88	2.35	2.01	417
$\rho_{12} = 0.5$							
$\rho_{13} = -0.5$	0.561	1.56	3.23	5.34	4.64	4.40	543
$\rho_{13} = 0$	0.500	1.08	1.73	3.08	2.62	2.40	542
$\rho_{13} = 0.5$	0.454	0.833	1.18	2.87	2.34	2.01	541
$\rho_{12} = -0.5$							
$\rho_{13} = -0.5$	0.565	1.59	3.27	5.39	4.66	4.42	571
$\rho_{13} = 0$	0.504	1.10	1.75	3.11	2.63	2.41	569
$\rho_{13} = 0.5$	0.457	0.842	1.20	2.90	2.36	2.02	569

Table 3.3. Certainty equivalents ($\times 10^{-3}$) for three possible futures combinations with two different maturities, different values of speed of reversion κ_1 and correlation ρ in the CTOU model.

	Futures combinations (maturity)		
Parameters	T_1	T_2	$\{T_1, T_2\}$
$\kappa_1 = 5.827$			
$\rho = -0.5$	3.66	45.6	209
$\rho = 0$	3.96	36.3	209
$\rho = 0.5$	2.48	19.4	209
$\kappa_1 = 0.827$			
$\rho = -0.5$	0.0281	0.290	209
$\rho = 0$	0.0393	0.377	209
$\rho = 0.5$	0.0264	0.259	209

that higher speed of mean reversion κ_1 leads to higher certainty equivalent. However, when two futures contracts are traded, the dependence of certainty equivalents on κ_1 and correlation parameter ρ disappears, which has been already shown by value function (3.24). This phenomenon is also pointed out by Leung and Yan (2018) that the certainty equivalent only depends on market prices of risk ζ_1 and ζ_2 when two futures are traded.

3.6 Conclusion

We have studied the optimal trading in commodity futures market under a multifactor model. Closed-form expressions for the optimal controls and for the value function are obtained. Using these results, the optimal futures trading strategies are analytically derived and numerically illustrated.

One novel feature of the portfolio optimization approach herein is that it allows different numbers of futures of with various maturities to be traded. Intuitively, it should be more valuable to the investor to access a larger set of securities, and this intuition is confirmed quantitatively through the certainty equivalents associated with the optimal futures portfolios.

In the presence of market frictions, trading more contracts might not necessarily be more beneficial. Therefore, it is practically important to determine the right number of contracts and rebalancing strategies accounting for transaction costs. Such an extension to our model would be interesting for future research.

Chapter 4

Futures Pricing in a Regime-Switching Market

Asset price dynamics may depend on market regimes, which can change suddenly and persist for a period of time. The timing of regime switching is often unpredictable, thus making hedging and risk management very challenging. In this chapter, we discuss a general regime-switching model to capture such properties of market dynamics. Specifically, an exogenous continuous-time finite-state Markov chain is used to represent the stochastic market regime.

Our regime-switching market approach represents a general framework that captures a number of regime-switching models. In this chapter, we analyze the futures pricing problem under this regime-switching framework, with applications to the regime-switching geometric Brownian motion (RS-GBM) and regime-switching exponential Ornstein–Uhlenbeck (RS-XOU) models. The no-arbitrage futures pricing formulas and price dynamics are derived. Our results reveal the impact of regime changes on the futures positions as well as portfolio returns.

The rest of this chapter is structured as follows. We describe the general market framework in Section 4.1. We apply our framework to the RS-GBM model and RS-XOU model in Sections 4.2 and 4.3, respectively. For all models presented herein, numerical implementation is discussed and illustrative examples are provided.

4.1 Futures Price Dynamics

We fix a probability space $(\Omega, \mathcal{G}, \mathbb{Q})$, where $\mathbb{Q}$ is the risk-neutral pricing measure $\mathbb{Q}$. Let ξ be a continuous-time irreducible finite-state Markov chain with state space $E = \{1, 2, \ldots, M\}$. Under the risk-neutral pricing measure, the generator matrix of ξ, denoted by $\tilde{\boldsymbol{Q}}$, has entries

$$\tilde{\boldsymbol{Q}}(i, j) = \tilde{q}(i, j),$$

such that entries satisfy

$$\begin{aligned} \tilde{q}(i, j) &\geq 0, \quad \text{for } i \neq j, \\ \sum_{j \in E} \tilde{q}(i, j) &= 0, \quad \text{for } i \in E. \end{aligned}$$

This Markov chain represents the changing market regime and influences the underlying asset's price dynamics. We illustrate an example of a two-state Markov chain in Figure 4.1.

We can use a stochastic integral with respect to a Poisson random measure to represent the Markov chain ξ. For $i, j \in E$ with $i \neq j$, let $\Delta(i, j)$ be the consecutive left-closed, right-open intervals of the real line, each having length $\tilde{q}(i, j)$. Define a function $h : E \times \mathbb{R} \to \mathbb{R}$ by

$$h(i, z) = \sum_{j \in E \setminus \{i\}} (j - i) I_{\{z \in \Delta(i, j)\}}.$$

Then, under measure $\mathbb{Q}$, the Markov chain ξ_t evolves according to

$$d\xi_t = \int_{\mathbb{R}} h(\xi_t, z) N(dt, dz),$$

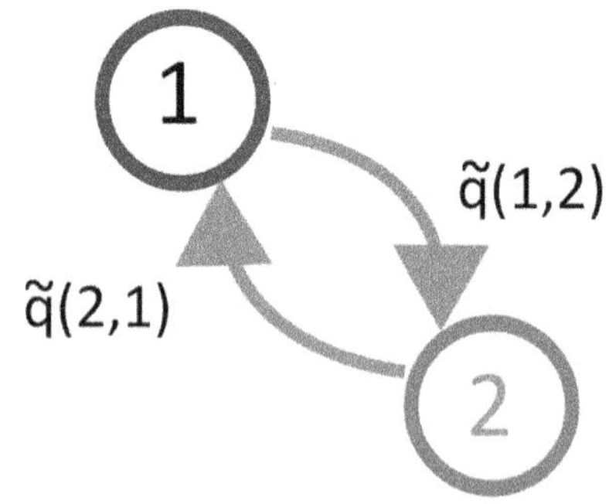

Fig. 4.1. Illustration of the Markov chain with two states $E = \{1, 2\}$.

where $N(dt, dz)$ is the Poisson random measure with intensity $dt \times \tilde{\mu}(dz)$ and $\tilde{\mu}$ is the Lebesgue measure satisfying

$$\int_{\mathbb{R}} I_{\{z \in \Delta(i,j)\}} \tilde{\mu}(dz) = |\Delta(i,j)| = \tilde{q}(i,j),$$

with $|\Delta(i,j)|$ being the length of $\Delta(i,j)$.

In turn, we can alternatively express the above stochastic differential equation (SDE) as

$$d\xi_t = \sum_{j \in E \backslash \{\xi_t\}} \tilde{q}(\xi_t, j)(j - \xi_t)dt + \int_{\mathbb{R}} h(\xi_t, z) M^{\mathbb{Q}}(dt, dz),$$

using the compensated Poisson process under measure $\mathbb{Q}$ defined by

$$M^{\mathbb{Q}}(dt, dz) = N(dt, dz) - dt \times \tilde{\mu}(dz).$$

The underlying asset's spot price is denoted by S_t. Its log-price is denoted by

$$X_t = \log(S_t),$$

and it evolves according to

$$dX_t = \tilde{a}(t, X_t, \xi_t)dt + b(t, X_t, \xi_t)dZ_t^{\mathbb{Q}}, \tag{4.1}$$

where $Z^{\mathbb{Q}}$ is the standard Brownian motion under the measure $\mathbb{Q}$ and independent of ξ. The drift and volatility functions $\tilde{a}(\cdot,\cdot,\cdot)$ and $b(\cdot,\cdot,\cdot)$ are assumed to satisfy the conditions such that SDE (4.1) has a strong solution.

Consider a futures contract on the underlying asset S with maturity T. Like (2.6) in Chapter 3, the no-arbitrage price of this futures at time $t \leq T$ is given by the conditional expectation under the risk-neutral pricing measure $\mathbb{Q}$:

$$F_i(t, x) = \mathbb{E}^{\mathbb{Q}}[\exp(X_T) | X_t = x, \xi_t = i].$$

The futures price function $F_i(t, x)$ is determined from the following system of PDEs

$$\partial_t F_i + \mathcal{L}_i^{\mathbb{Q}} F_i + \sum_{j \in E \backslash \{i\}} \tilde{q}(i,j)(F_j - F_i) = 0, \tag{4.2}$$

for $(t,x) \in [0,T) \times \mathbb{R}$ and $i = 1, \ldots, M$, where

$$\mathcal{L}_i^{\mathbb{Q}} \cdot := \tilde{a}(t,x,i)\partial_x \cdot + \frac{b^2(t,x,i)}{2}\partial_{xx} \cdot$$

To facilitate presentation, we have dropped the variables from different functions in (4.2) and will do the same in PDEs that follow when no ambiguity arises.

For the futures trading problem, the underlying asset and futures prices are observed under the physical measure $\mathbb{P}$. Under measure $\mathbb{P}$, the Markov chain ξ has generator matrix $\boldsymbol{Q}$ with entries $\boldsymbol{Q}(i,j) = q(i,j)$, with $i, j \in E$. Since $\mathbb{P}$ and $\mathbb{Q}$ are equivalent measures, we have

$$q(i,j) = 0 \quad \text{iff } \widetilde{q}(i,j) = 0.$$

To relate the Poisson random measures under measures $\mathbb{P}$ and $\mathbb{Q}$, we denote by $\mu(dz)$ the intensity measure of $N(dt,dz)$ under measure $\mathbb{P}$ such that

$$\mu(dz) = \begin{cases} \dfrac{q(i,j)}{\widetilde{q}(i,j)}\widetilde{\mu}(dz), & \text{for } z \in \Delta(i,j), \\ \widetilde{\mu}(dz), & \text{others,} \end{cases}$$

under the convention that $0/0 = 1$.

As a consequence, the compensated Poisson process under measure $\mathbb{P}$ is

$$M^{\mathbb{P}}(dt,dz) = M^{\mathbb{Q}}(dt,dz) - \sum_{i,j \in E, i \neq j} \frac{q(i,j) - \widetilde{q}(i,j)}{\widetilde{q}(i,j)} I_{\{z \in \Delta(i,j)\}} dt \times \widetilde{\mu}(dz).$$

Accordingly, the Markov chain ξ_t satisfies

$$d\xi_t = \sum_{j \in E \setminus \{\xi_t\}} q(\xi_t, j)(j - \xi_t)dt + \int_{\mathbb{R}} h(\xi_t, z) M^{\mathbb{P}}(dt,dz).$$

To relate the Brownian motions under measures $\mathbb{P}$ and $\mathbb{Q}$, we denote by $\zeta(\xi_t)$ the risk premium associated with the Brownian motion such that

$$dZ_t^{\mathbb{Q}} = dZ_t^{\mathbb{P}} + \zeta(\xi_t)dt. \tag{4.3}$$

In turn, the log spot price satisfies

$$dX_t = a(t, X_t, \xi_t)dt + b(t, X_t, \xi_t)dZ_t^{\mathbb{P}},$$

whose drift is given by

$$a(t,x,i) := \widetilde{a}(t,x,i) + \zeta(i)b(t,x,i).$$

Here, $Z^{\mathbb{P}}$ is the standard Brownian motion under $\mathbb{P}$ and is independent of ξ.

Applying Itô's lemma, the futures price F_t satisfies

$$dF_t = \eta(t,X_t,\xi_t)dZ_t^{\mathbb{Q}} + \int_{\mathbb{R}} \sum_{j\in E\setminus\{\xi_t\}} \Delta_F(t,X_t,\xi_t,j) I_{\{z\in\Delta(\xi_t,j)\}} M^{\mathbb{Q}}(dt,dz), \tag{4.4}$$

where we have defined

$$\begin{aligned} \eta(t,x,i) &= b(t,x,i)\partial_x F_i(t,x), \\ \Delta_F(t,x,i,j) &= F_j(t,x) - F_i(t,x), \end{aligned} \tag{4.5}$$

for $i, j \in E$. In particular, we have

$$\Delta_F(t,x,i,i) = 0$$

by definition. In addition, we note that F_t is a $\mathbb{Q}$-martingale.

Next, applying the risk premium equation (4.3), we obtain the $\mathbb{P}$-dynamics of the futures price F_t, described by the SDE:

$$\begin{aligned} dF_t = &\left(\eta(t,X_t,\xi_t)\zeta(\xi_t) + \sum_{j\in E\setminus\{\xi_t\}} (q(\xi_t,j) - \widetilde{q}(\xi_t,j))\Delta_F(t,X_t,\xi_t,j)\right)dt \\ &+ \eta(t,X_t,\xi_t)dZ_t^{\mathbb{P}} + \int_{\mathbb{R}} \sum_{j\in E\setminus\{\xi_t\}} \Delta_F(t,X_t,\xi_t,j) I_{\{z\in\Delta(\xi_t,j)\}} M^{\mathbb{P}}(dt,dz). \end{aligned}$$

A key feature of this regime-switching framework is that the futures price process is a jump-diffusion even though the spot price process has continuous paths.

4.2 Regime-Switching Geometric Brownian Motion

Suppose the log-price of the underlying asset follows the SDE:

$$dX_t = \mu(\xi_t)dt + \sigma(\xi_t)dZ_t^{\mathbb{Q}},$$

under the risk-neutral measure $\mathbb{Q}$. We call this model the RS-GBM because without ξ the spot price S is simply a GBM. This model belongs to our regime-switching framework discussed in the previous section. Indeed, this amounts to setting the coefficients in SDE (4.1) as follows:

$$\widetilde{a}(t, X_t, \xi_t) = \mu(\xi_t), \quad \text{and} \quad b(t, X_t, \xi_t) = \sigma(\xi_t). \tag{4.6}$$

Substituting (4.6) into (4.2), we obtain the PDE system for the futures price function under this model. Precisely, for $i \in E$, we have

$$\partial_t F_i + \mu_i \partial_x F_i + \frac{\sigma_i^2}{2} \partial_{xx} F_i + \sum_{j \in E \setminus \{i\}} \widetilde{q}(i,j)(F_j - F_i) = 0,$$

with

$$\mu_i = \mu(i), \quad \text{and} \quad \sigma_i = \sigma(i).$$

The terminal condition is

$$F_i(T, x) = e^x, \quad x \in \mathbb{R}.$$

Under this model, the futures price admits the separation of variables:

$$F_i(t, x) = e^x \, g_i(t),$$

where $(g_i(t))_{i=1,\dots,M}$ solve the system of ODEs:

$$\frac{dg_i(t)}{dt} + \left(\mu_i + \frac{\sigma_i^2}{2}\right) g_i(t) + \sum_{j \in E \setminus \{i\}} \widetilde{q}(i,j)(g_j(t) - g_i(t)) = 0, \tag{4.7}$$

for $t \in [0, T)$, with the terminal condition $g_i(T) = 1$, for $i = 1, \dots, M$. Defining $\boldsymbol{g}(t) = (g_1(t), \dots, g_M(t))^\top$, we can write the solution as

$$\boldsymbol{g}(t) = \exp\left((\boldsymbol{G} + \widetilde{\boldsymbol{Q}})(T - t)\right)\mathbf{1},$$

where $\widetilde{\boldsymbol{Q}}$ is the generator matrix under the measure $\mathbb{Q}$ and

$$\boldsymbol{G} = \text{diag}\left(\frac{2\mu_1 + \sigma_1^2}{2}, \frac{2\mu_2 + \sigma_2^2}{2}, \dots, \frac{2\mu_M + \sigma_M^2}{2}\right).$$

In addition, the ODE system (4.7) implies a probabilistic representation for $g_i(t)$:

$$g_i(t) = \mathbb{E}^{\mathbb{Q}}\left[\exp\left(\int_t^T \mu(\xi_s) + \frac{\sigma^2(\xi_s)}{2} ds\right) \middle| \xi_t = i\right]. \tag{4.8}$$

To sum up, the futures price in regime i is given by

$$F_i(t,x) = \exp(x)\big(\exp((\boldsymbol{G} + \widetilde{\boldsymbol{Q}})(T-t))\mathbf{1}\big)_i, \tag{4.9}$$

where the subscript i denotes the ith entry of the vector.

In turn, using (4.4), the futures price satisfies the SDE

$$\begin{aligned} dF_t = &\Bigg(\sigma(\xi_t)F(t,X_t,\xi_t)\zeta(\xi_t) \\ &+ \sum_{j\in E\backslash\{\xi_t\}} (q(\xi_t,j) - \widetilde{q}(\xi_t,j))(F(t,X_t,j) - F(t,X_t,\xi_t))\Bigg)dt \\ &+ \sigma(\xi_t)F(t,X_t,\xi_t)dZ_t^{\mathbb{P}} \\ &+ \int_{\mathbb{R}} \sum_{j\in E\backslash\{\xi_t\}} (F(t,X_t,j) - F(t,X_t,\xi_t))I_{\{z\in\Delta(\xi_t,j)\}}M^{\mathbb{P}}(dt,dz). \end{aligned} \tag{4.10}$$

To illustrate the futures trading problem under the RS-GBM model, we simulate the sample paths for spot price, futures price, optimal investment, and optimal wealth process. The regime switching between two states, with transition probabilities $q_{12} = 2$ and $q_{21} = 4$ which are entries of generator matrix $\boldsymbol{Q}$. The trading horizon $\widetilde{T} = 0.6$ which is no greater than the maturity of the futures contract $T_1 = 0.6$ and $T_2 = 0.8$. All parameters are summarized in Table 4.1.

As shown in Figure 4.2, the market starts in regime 2, then switches to regime 1 at time t_1 before returning to regime 2 at time t_2. Under RS-GBM model, the futures price, given by formula (4.9), tends to amplify the spot price. Thus, the futures price is volatile relative to the spot price. Moreover, each regime switch can cause an instant jump in the futures price but not in the spot price. Since the trading horizon coincides with the maturity of the T_1-futures, the corresponding futures price converges to the spot price toward the end.

Table 4.1. Parameters for the RS-GBM model for Figures 4.2–5.4.

$\widetilde{q}_{12} = q_{12}$	$\widetilde{q}_{21} = q_{21}$	$\zeta(1)$	$\zeta(2)$	γ	$\widetilde{T}$
2	4	0.1	0.3	1	0.6
μ_1	μ_2	σ_1	σ_2	T_1	T_2
−0.2	0.2	0.2	0.3	0.6	0.8

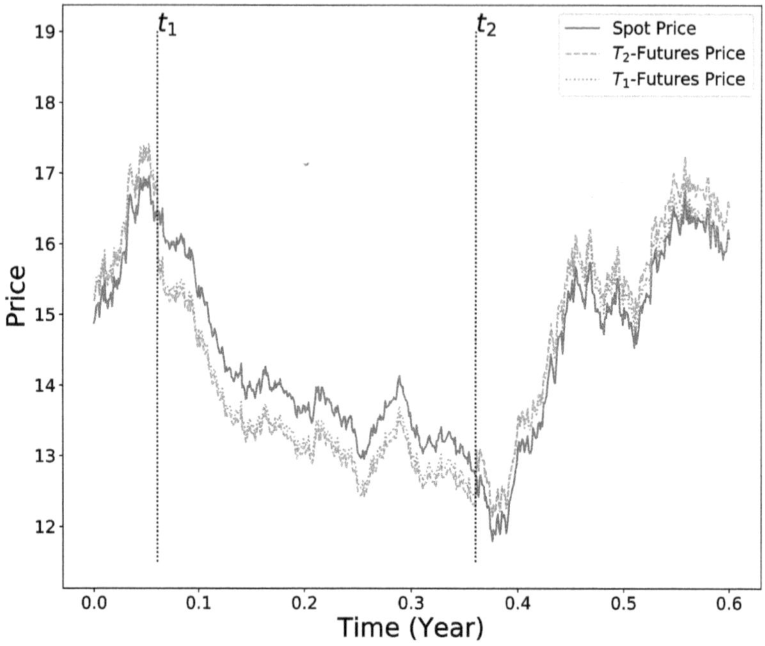

Fig. 4.2. Sample paths of the spot price, T_1-futures price, and T_2-futures price over the trading horizon under the RS-GBM model. The market starts in regime 2, then switches to regime 1 at time t_1, before switching back to regime 2 at time t_2.

4.3 Regime-Switching Exponential Ornstein–Uhlenbeck Model

As is well known, the exponential Ornstein–Uhlenbeck process and its variations are widely used to model commodity prices. We now consider the RS-XOU model and illustrate the optimal trading strategies under this model. In the RS-XOU model, the log spot price evolves according to

$$dX_t = \kappa(\xi_t)(\theta(\xi_t) - X_t)dt + \sigma(\xi_t)dZ_t^{\mathbb{Q}}, \tag{4.11}$$

where $\kappa(\xi_t)$, $\theta(\xi_t)$ and $\sigma(\xi_t)$ are the functions of regimes. This amounts to setting

$$\begin{aligned} \tilde{a}(t, X_t, \xi_t) &= \kappa(\xi_t)(\theta(\xi_t) - X_t), \\ b(t, X_t, \xi_t) &= \sigma(\xi_t) \end{aligned}$$

in (4.1).

4.3.1 Futures dynamics

The futures price function $F_i(t,x)$ satisfies PDE (4.2). Substituting (4.11) into (4.4), we obtain the $\mathbb{Q}$-dynamics for futures price F_t:

$$\begin{aligned}dF_t &= \eta(t,X_t,\xi_t)dZ_t^{\mathbb{Q}}\\ &\quad + \int_{\mathbb{R}} \sum_{j\in E\setminus\{\xi_t\}} (F(t,X_t,j)-F(t,X_t,\xi_t))I_{\{z\in\Delta(\xi_t,j)\}}M^{\mathbb{Q}}(dt,dz),\end{aligned}$$

where the volatility term is given by

$$\eta(t,x,i)=\sigma_i\partial_x F_i(t,x).$$

In addition, under the measure $\mathbb{P}$,

$$\begin{aligned}dF_t &= \Bigg(\varsigma(\xi_t)\eta(t,X_t,\xi_t)\\ &\quad + \sum_{j\in E\setminus\{\xi_t\}} (q(\xi_t,j)-\widetilde{q}(\xi_t,j))(F(t,X_t,j)-F(t,X_t,\xi_t))\Bigg)dt\\ &\quad + \eta(t,X_t,\xi_t)dZ_t^{\mathbb{P}} + \int_{\mathbb{R}} \sum_{j\in E\setminus\{\xi_t\}} (F(t,X_t,j)\\ &\quad - F(t,X_t,\xi_t))I_{\{z\in\Delta(\xi_t,j)\}}M^{\mathbb{P}}(dt,dz).\end{aligned}$$

4.3.2 Numerical implementation and examples

To apply the finite difference method, we first re-write futures price PDE (4.2) in terms of $s=e^x$ to get

$$\partial_t F_i + \left(\kappa_i(\theta_i-\ln s)+\frac{\sigma_i^2}{2}\right)s\partial_s F_i + \frac{\sigma_i^2 s^2}{2}\partial_{ss}F_i + \sum_{j\in E\setminus\{i\}} \widetilde{q}(i,j)(F_j-F_i)=0, \tag{4.12}$$

for $(t,s)\in[0,T)\times\mathbb{R}_+$, with the terminal condition

$$F_i(T,s)=s.$$

Then we can apply the Crank–Nicolson method to equation (4.12) directly with Dirichlet boundary conditions. The method is standard, so we omit the details here.

Next, we discuss the Fourier time-stepping method. begin first by discretizing the continuous-time Markov chain ξ_t with time step of size δt, and we keep ξ_t constant on each time interval $(t_n, t_{n+1}]$ $(t_n = n\delta t)$, for $n = 0, \ldots, T/\delta t - 1$, with transition probabilities

$$P_{kl} := \begin{cases} 1 + \widetilde{q}_{ll}\delta t, & k = l, \\ \widetilde{q}_{kl}\delta t, & \text{otherwise.} \end{cases}$$

In turn, the futures price satisfies the recursive relation

$$F_i(t_n, x) = \sum_{j=1,\ldots,M} P_{ij} F_j(t_{n+}, x), \tag{4.13}$$

where

$$F_j(t_{n+}, x) = \lim_{t \downarrow t_n} F_j(t, x).$$

Since we assume Markov chain ξ_t stay constant on time intervals $(t_n, t_{n+1}]$, the martingale property of futures price implies that

$$F_i(t_{n+}, x) = \mathbb{E}^{\mathbb{Q}}[F_i(t_{n+1}, X_{t_{n+1}}) | X_{t_n} = x],$$

for $i = 1, \ldots, M$, and $n = 0, \ldots, T/\delta t - 1$. Therefore, we have following PDE for the futures price within each time interval $(t_n, t_{n+1}]$ and regime i,

$$\partial_t F_i + \kappa_i(\theta_i - x)\partial_x F_i + \frac{\sigma_i^2}{2}\partial_{xx} F_i = 0. \tag{4.14}$$

Next, we apply Fourier transform to PDE (4.14). To that end, we define

$$\mathcal{F}[f](\omega) = \int_{-\infty}^{\infty} f(x)\exp(-j\omega x)dx,$$

where j denotes the imaginary identity and ω denotes the frequency. Then, we obtain a first-order PDE for $\hat{F} := \mathcal{F}[F]$:

$$\partial_t \hat{F}_i + \kappa_i \omega \partial_\omega \hat{F}_i + \left(\kappa_i \theta_i j\omega + \kappa_i - \frac{\omega^2 \sigma_i^2}{2}\right)\hat{F}_i = 0, \tag{4.15}$$

for $(t, \omega) \in [0, T) \times \mathbb{R}$, where

$$\hat{F}(t, \omega) \equiv \mathcal{F}[F](t, \omega).$$

Noted that we use the following property of the Fourier transform to derive PDE (4.15),

$$\mathcal{F}[xf_x] = -\mathcal{F}[f] - \omega\mathcal{F}_\omega[f].$$

To solve PDE (4.15) we employ the method of characteristics to get

$$\hat{F}_i(t_n+,\omega) = \phi_i(\delta t,\omega)\hat{F}_i(t_{n+1}, e^{\kappa_i \delta t}\omega), \tag{4.16}$$

with

$$\phi_i(\delta t,\omega) = \exp\left(\kappa_i\delta t - \theta_i j\omega(1-e^{\kappa_i\delta t}) + \frac{\sigma_i^2\omega^2}{4\kappa_i}(1-e^{2\kappa_i\delta t})\right).$$

Combining equation (4.13) and (4.16), we have following backward equation in the frequency space:

$$\hat{F}_i(t_n,\omega) = \sum_{j=1,\dots,M} P_{ij}\phi_j(\delta t,\omega)\hat{F}_j(t_{n+1}, e^{\kappa_j\delta t}\omega), \tag{4.17}$$

for $i = 1,\dots,M$, and $n = 0,\dots,T/\delta t - 1$. We then apply backward induction via (4.17) to calculate $\hat{F}_i(t,\omega)$. To recover the original futures price function, we apply inverse Fourier transform. For the numerical implementation of Fourier transform and inverse Fourier transform, we utilize the standard fast Fourier transform algorithm. This FST method has been applied more broadly by Jackson *et al.* (2008) to solve partial-integro differential equations (PIDEs) that arise in options pricing problems. For more details and other applications of this numerical approach, we refer to Surkov (2009) and Jackson *et al.* (2008), and references therein. In the literature, Leung and Wan (2015) apply a Fourier time-stepping (FST) method to compute the cost of an American-style ESO when the company stock is driven by a Lévy process.

We simulate the sample paths for spot price, futures prices, optimal investment, and optimal wealth process. The regime switching between two states, with transition probabilities $q_{12} = 2$ and $q_{21} = 4$ which are entries of generator matrix $\boldsymbol{Q}$. The trading horizon $\widetilde{T} = 0.6$, and the two futures contracts have maturities $T_1 = 0.6$ and $T_2 = 0.8$. The parameters are summarized in Table 4.2. We will use these parameters to generate the numerical examples in Figures 4.3–5.8.

Table 4.2. Parameters for the RS-XOU model.

$\widetilde{q}_{12} = q_{12}$	$\widetilde{q}_{21} = q_{21}$	$\zeta(1)$	$\zeta(2)$	κ_1	κ_2	γ
2	4	0.1	0.3	1	2	1
θ_1	θ_2	σ_1	σ_2	T_1	T_2	$\widetilde{T}$
2.5	2.7	0.2	0.3	0.6	0.8	0.6

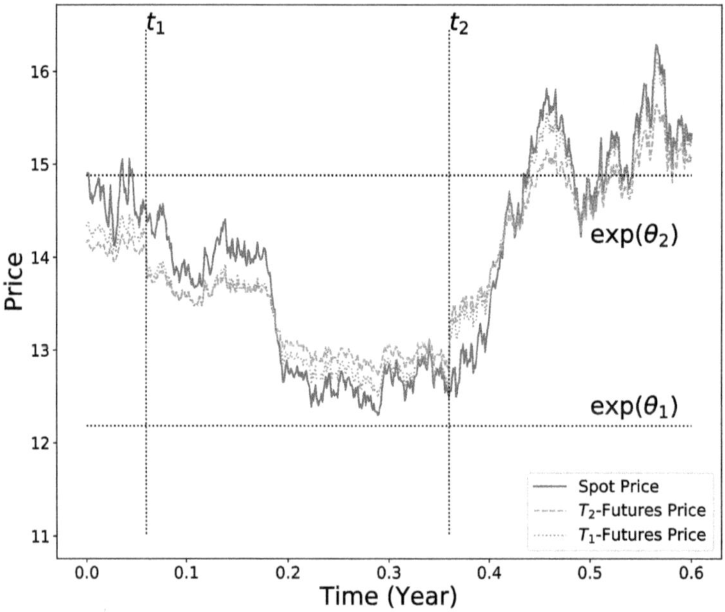

Fig. 4.3. Sample paths of the spot price, T_1-futures price, and T_2-futures price over the trading horizon under the RS-XOU model. The market starts in regime 2, then switches to regime 1 at time t_1, before switching back to regime 2 at time t_2.

For the sample paths shown in Figure 4.3, the market starts in regime 2, then switches to regime 1 at time t_1) before returning to regime 2 at time t_2. The two price levels $\exp(\theta_1)$ and $\exp(\theta_2)$ are the long-run means of the spot price in regimes 1 and 2, respectively. In each regime, the spot price tends to move towards the corresponding mean level. The spot and futures prices tend to move in tandem. Nevertheless, like in Figure 4.2, each regime switch can cause an instant jump in the futures price but not in the spot price.

Chapter 5

Dynamic Futures Portfolio in a Regime-Switching Market

5.1 Introduction

In this chapter, we present a stochastic control approach to generate dynamic futures trading strategies under the regime-switching framework discussed in the previous chapter. We determine the optimal futures trading strategy by solving a utility maximization problem. By analyzing and solving the associated Hamilton–Jacobi–Bellman (HJB) equations, we derive the investor's value function and optimal trading strategies.

Among our findings, we show that the investor's value function admits a separable form under a general regime-switching market framework, and the original HJB equations are reduced to a system of linear ODEs. In addition, we also define the investor's certainty equivalent to quantify the value of the futures trading opportunity to the investor. Surprisingly both the value function and certainty equivalent do not depend on the current spot and futures prices, and they admit universal forms across different model specifications. Nevertheless, the risk premia associated with the regime-switching model play a crucial role, not only in the value function and certainty equivalent, but also in the optimal strategy. In addition, we show two applications of our model, regime-switching geometric Brownian motion (RS-GBM) and regime-switching exponential Ornstein–Uhlenbeck (RS-XOU) model.

The rest of this chapter is structured as follows. We describe the general market framework in Section 5.2. The dynamic futures portfolio optimization is discussed in Section 5.3. We apply our framework to the RS-GBM model and RS-XOU model in Sections 5.4 and 5.5, respectively. Concluding remarks are provided in Section 5.6.

5.2 Futures Price Dynamics

As in Chapter 4, we fix a probability space $(\Omega, \mathcal{G}, \mathbb{Q})$, where $\mathbb{Q}$ is the risk-neutral pricing measure $\mathbb{Q}$. Let ξ be a continuous-time irreducible finite-state Markov chain with state space $E = \{1, 2, \ldots, M\}$. The generator matrix of ξ, denoted by $\tilde{\boldsymbol{Q}}$, has entries $\tilde{\boldsymbol{Q}}(i,j) = \tilde{q}(i,j)$ such that $\tilde{q}(i,j) \geq 0$ for $i \neq j$ and $\sum_{j\in E} \tilde{q}(i,j) = 0$ for $i \in E$. This Markov chain represents the changing market regime and influences the underlying asset's price dynamics.

We can use a stochastic integral with respect to a Poisson random measure to represent Markov chain ξ. For $i, j \in E$ with $i \neq j$, let $\Delta(i,j)$ be the consecutive left-closed, right-open intervals of the real line, each having length $\tilde{q}(i,j)$. Define a function $h : E \times \mathbb{R} \to \mathbb{R}$ by

$$h(i,z) = \sum_{j\in E\setminus\{i\}} (j-i) I_{\{z\in\Delta(i,j)\}}.$$

Then, under measure $\mathbb{Q}$, the Markov chain ξ_t evolves according to

$$d\xi_t = \int_{\mathbb{R}} h(\xi_t, z) N(dt, dz),$$

where $N(dt, dz)$ is the Poisson random measure with intensity $dt \times \tilde{\mu}(dz)$ and $\tilde{\mu}$ is the Lebesgue measure satisfying

$$\int_{\mathbb{R}} I_{\{z\in\Delta(i,j)\}} \tilde{\mu}(dz) = |\Delta(i,j)| = \tilde{q}(i,j),$$

with $|\Delta(i,j)|$ being the length of $\Delta(i,j)$. The above SDE can be also be written as

$$d\xi_t = \sum_{j\in E\setminus\{\xi_t\}} \tilde{q}(\xi_t, j)(j-\xi_t)dt + \int_{\mathbb{R}} h(\xi_t, z) M^{\mathbb{Q}}(dt, dz),$$

using the compensated Poisson process under measure $\mathbb{Q}$ defined by

$$M^{\mathbb{Q}}(dt, dz) = N(dt, dz) - dt \times \tilde{\mu}(dz).$$

The underlying asset's spot price is denoted by S_t. Its log-price, $X_t = \log(S_t)$, evolves according to

$$dX_t = \tilde{a}(t, X_t, \xi_t)dt + b(t, X_t, \xi_t)dZ_t^{\mathbb{Q}},$$

where $Z^{\mathbb{Q}}$ is the standard Brownian motion under the measure $\mathbb{Q}$ and independent of ξ, and $\tilde{a}(\cdot,\cdot,\cdot)$ and $b(\cdot,\cdot,\cdot)$ are the drift and volatility functions.

Under this framework, the no-arbitrage price of a futures contract on the underlying asset S with maturity T is given by the conditional expectation under the risk-neutral pricing measure $\mathbb{Q}$:

$$F_i(t,x) = I\!E^{\mathbb{Q}}[\exp(X_T)|X_t = x, \xi_t = i].$$

The futures price function $F_i(t,x)$ is determined from the following system of PDEs

$$\partial_t F_i + \mathcal{L}_i^{\mathbb{Q}} F_i + \sum_{j\in E\setminus\{i\}} \tilde{q}(i,j)(F_j - F_i) = 0, \tag{5.1}$$

for $(t,x) \in [0,T) \times \mathbb{R}$ and $i = 1,\ldots,M$, where

$$\mathcal{L}_i^{\mathbb{Q}} \cdot := \tilde{a}(t,x,i)\partial_x \cdot + \frac{b^2(t,x,i)}{2}\partial_{xx} \cdot \, .$$

To facilitate presentation, we have dropped the variables from different functions in (5.1) and will do the same in PDEs that follow when no ambiguity arises.

For the futures trading problem, asset and futures prices are observed under the physical measure $\mathbb{P}$. Under measure $\mathbb{P}$, the Markov chain ξ has generator matrix $\boldsymbol{Q}$ with entries $\boldsymbol{Q}(i,j) = q(i,j)$, where $i,j \in E$. Since $\mathbb{P}$ and $\mathbb{Q}$ are equivalent measures, we have $q(i,j) = 0$ iff $\widetilde{q}(i,j) = 0$. To relate the Poisson random measures under measures $\mathbb{P}$ and $\mathbb{Q}$, we denote by $\mu(dz)$ the intensity measure of $N(dt,dz)$ under measure $\mathbb{P}$ such that

$$\mu(dz) = \begin{cases} \dfrac{q(i,j)}{\widetilde{q}(i,j)}\widetilde{\mu}(dz), & \text{for } z \in \Delta(i,j), \\ \widetilde{\mu}(dz), & \text{others}, \end{cases}$$

under the convention that $0/0 = 1$. Then, the compensated Poisson process under measure $\mathbb{P}$ is

$$M^{\mathbb{P}}(dt,dz) = M^{\mathbb{Q}}(dt,dz) - \sum_{i,j\in E, i\neq j} \frac{q(i,j) - \widetilde{q}(i,j)}{\widetilde{q}(i,j)} I_{\{z\in\Delta(i,j)\}} dt \times \widetilde{\mu}(dz).$$

Accordingly, the Markov chain ξ_t satisfies

$$d\xi_t = \sum_{j\in E\setminus\{\xi_t\}} q(\xi_t,j)(j - \xi_t)dt + \int_{\mathbb{R}} h(\xi_t,z) M^{\mathbb{P}}(dt,dz).$$

To connect the Brownian motions under measures $\mathbb{P}$ and $\mathbb{Q}$, we denote by $\zeta(\xi_t)$ the risk premium associated with the Brownian motion such that

$$dZ_t^{\mathbb{Q}} = dZ_t^{\mathbb{P}} + \zeta(\xi_t)dt.$$

Then, the log spot price satisfies

$$dX_t = a(t, X_t, \xi_t)dt + b(t, X_t, \xi_t)dZ_t^{\mathbb{P}},$$

whose drift is given by

$$a(t, x, i) := \tilde{a}(t, x, i) + \zeta(i)b(t, x, i).$$

Here, $Z^{\mathbb{P}}$ is the standard Brownian motion under $\mathbb{P}$ and is independent of ξ.

The futures price F_t satisfies the SDE

$$dF_t = \eta(t, X_t, \xi_t)dZ_t^{\mathbb{Q}} + \int_{\mathbb{R}} \sum_{j\in E\setminus\{\xi_t\}} \Delta_F(t, X_t, \xi_t, j) I_{\{z\in\Delta(\xi_t,j)\}} M^{\mathbb{Q}}(dt, dz),$$

with

$$\eta(t, x, i) = b(t, x, i)\partial_x F_i(t, x),$$
$$\Delta_F(t, x, i, j) = F_j(t, x) - F_i(t, x),$$

for $i, j \in E$. In particular, $\Delta_F(t, x, i, i) = 0$ by definition. Since F_t is a $\mathbb{Q}$-martingale, after a change of measure via (4.3), the futures price F_t admits the $\mathbb{P}$-dynamics:

$$dF_t = \Bigg(\eta(t, X_t, \xi_t)\zeta(\xi_t) + \sum_{j\in E\setminus\{\xi_t\}} (q(\xi_t, j) - \tilde{q}(\xi_t, j))\Delta_F(t, X_t, \xi_t, j)\Bigg)dt$$
$$+ \eta(t, X_t, \xi_t)dZ_t^{\mathbb{P}} + \int_{\mathbb{R}} \sum_{j\in E\setminus\{\xi_t\}} \Delta_F(t, X_t, \xi_t, j) I_{\{z\in\Delta(\xi_t,j)\}} M^{\mathbb{P}}(dt, dz).$$

Even though the spot price process has continuous paths, the futures price process is a jump-diffusion. The jumps in the futures prices will have direct impact on the strategies in dynamic futures portfolio.

5.3 Futures Portfolio Optimization

We now consider the problem of dynamic trading futures contracts. Consider M futures $\boldsymbol{F} = (F^{(1)}, \ldots, F^{(M)})$ written on the same asset S with different maturities, denoted by $T_1 < T_2 < \cdots < T_M$ without loss of generality.

We define $M \times M$ *coefficient matrix* by

$$\mathbf{\Gamma}(t,x,i) = \begin{bmatrix} \eta^{(1)}(t,x,i) & \eta^{(2)}(t,x,i) & \cdots & \eta^{(M)}(t,x,i) \\ \Delta_F^{(1)}(t,x,i,1) & \Delta_F^{(2)}(t,x,i,1) & \cdots & \Delta_F^{(M)}(t,x,i,1) \\ \vdots & \vdots & \vdots & \vdots \\ \Delta_F^{(1)}(t,x,i,i-1) & \Delta_F^{(2)}(t,x,i,i-1) & \cdots & \Delta_F^{(M)}(t,x,i,i-1) \\ \Delta_F^{(1)}(t,x,i,i+1) & \Delta_F^{(2)}(t,x,i,i+1) & \cdots & \Delta_F^{(M)}(t,x,i,i+1) \\ \vdots & \vdots & \ddots & \vdots \\ \Delta_F^{(1)}(t,x,i,M) & \Delta_F^{(2)}(t,x,i,M) & \cdots & \Delta_F^{(M)}(t,x,i,M) \end{bmatrix}. \tag{5.2}$$

The intuition here is that the coefficient matrix gives a link between the traded futures and sources of randomness:

$$\begin{bmatrix} dF_t^{(1)} \\ dF_t^{(2)} \\ \cdots \\ dF_t^{(M)} \end{bmatrix} \longleftrightarrow \begin{bmatrix} dZ_t^{\mathbb{Q}} \\ \int_{\mathbb{R}} I_{\{z\in\Delta(\xi_t,j)\}} M^{\mathbb{Q}}(dt,dz) \\ \cdots \\ \int_{\mathbb{R}} I_{\{z\in\Delta(\xi_t,\xi_t-1)\}} M^{\mathbb{Q}}(dt,dz) \\ \int_{\mathbb{R}} I_{\{z\in\Delta(\xi_t,\xi_t+1)\}} M^{\mathbb{Q}}(dt,dz) \\ \cdots \\ \int_{\mathbb{R}} I_{\{z\in\Delta(\xi_t,M)\}} M^{\mathbb{Q}}(dt,dz) \end{bmatrix}.$$

It follows from (4.4) that if this $M \times M$ matrix $\mathbf{\Gamma}$ is invertible, then it is sufficient to these M futures to fully replicate any other futures on S with a different maturity, up to the shortest maturity. To see this, for a futures with an arbitrary maturity T, its price is connected with the prices of the M futures as follows:

$$dF_t = \begin{bmatrix} dF_t^{(1)} & dF_t^{(2)} & \cdots & dF_t^{(M)} \end{bmatrix} \mathbf{\Gamma}(t,X_t,\xi_t)^{-1} \begin{bmatrix} \eta(t,X_t,\xi_t) \\ \Delta_F(t,X_t,\xi_t,1) \\ \vdots \\ \Delta_F(t,X_t,\xi_t,\xi_t-1) \\ \Delta_F(t,X_t,\xi_t,\xi_t+1) \\ \vdots \\ \Delta_F(t,X_t,\xi_t,M) \end{bmatrix},$$

for $0 \le t \le T \wedge T_1$. In other words, the T-futures is redundant in this market with M regimes.

5.3.1 *Utility maximization*

We consider a portfolio of M futures with different maturities $T_1 < T_2 < \cdots < T_M$. We assume that these futures are not redundant, and the interest rate is zero. We fix the trading horizon $\widetilde{T}$, which must not exceed the shortest maturities of the futures in the portfolio. Hence, we require that $\widetilde{T} \le T_1$.

Like (3.5) in Chapter 3, we let strategy $\boldsymbol{\pi}_t = \left(\pi_t^{(1)}, \ldots, \pi_t^{(M)}\right)^\top$, where the element $\pi_t^{(k)}$ denotes the amount of money invested in kth futures contract. Then, the wealth process is given by

$$\begin{aligned} dW_t^{\varpi} &= \sum_{k=1}^{M} \pi_t^{(k)} \frac{dF_t^{(k)}}{F_t^{(k)}} \\ &= \sum_{k=1}^{M} \varpi_t^{(k)} \, dF_t^{(k)}, \end{aligned} \tag{5.3}$$

where $\varpi_t^{(k)} = \pi_t^{(k)} / F_t^{(k)}$ represents the units of futures $F^{(k)}$ in the portfolio.

We now reorganize terms in wealth process (5.3). To that end, we define the transformed strategies by

$$\widetilde{\varpi}_t^{(0)} = \sum_{k=1}^{M} \varpi_t^{(k)} \eta^{(k)}(t, X_t, \xi_t), \tag{5.4}$$

$$\widetilde{\varpi}_t^{(j)} = \sum_{k=1}^{M} \varpi_t^{(k)} \Delta_F^{(k)}(t, X_t, \xi_t, j), \quad \text{for } j \in E. \tag{5.5}$$

In particular, since $\Delta_F^{(k)}(t, x, i, i) = 0$, for $\forall k = 1, \ldots, M$ and $\forall i \in E$, according to (4.5), it follows that

$$\widetilde{\varpi}_t^{(\xi_t)} = \sum_{k=1}^{M} \varpi_t^{(k)} \Delta_F^{(k)}(t, X_t, \xi_t, \xi_t) = 0.$$

In matrix form, we have

$$\widetilde{\boldsymbol{\varpi}}_t = \boldsymbol{\Gamma}(t, X_t, \xi_t) \boldsymbol{\varpi}_t, \tag{5.6}$$

where $\widetilde{\boldsymbol{\varpi}}_t = (\widetilde{\varpi}_t^{(0)}, \widetilde{\varpi}_t^{(1)}, \ldots, \widetilde{\varpi}_t^{(\xi_t-1)}, \widetilde{\varpi}_t^{(\xi_t+1)}, \ldots, \widetilde{\varpi}_t^{(M)})^\top$. Applying (4.4), (5.4), and (5.5) to (5.3), the wealth process becomes

$$dW_t^{\varpi} = \left(\zeta(\xi_t)\widetilde{\varpi}_t^{(0)} + \sum_{j\in E\setminus\{\xi_t\}} (q(\xi_t,j) - \widetilde{q}(\xi_t,j))\widetilde{\varpi}_t^{(j)}\right)dt$$
$$+ \widetilde{\varpi}_t^{(0)} dZ_t^{\mathbb{P}} + \int_{\mathbb{R}} \sum_{j\in E\setminus\{\xi_t\}} \widetilde{\varpi}_t^{(j)} I_{\{z\in\Delta(\xi_t,j)\}} M^{\mathbb{P}}(dt,dz). \quad (5.7)$$

Note that the wealth process is subject to jumps whenever the market regime switches states.

We consider a utility maximization approach to determine the optimal futures trading strategy. The investor seeks to maximize the expected utility by dynamically trading the futures continuously over time. We assume the associated coefficient matrix $\boldsymbol{\Gamma}(t, X_t, \xi_t)$ be invertible, and it would act as a bijection mapping between $\boldsymbol{\varpi}_t$ and $\widetilde{\boldsymbol{\varpi}}_t$. This allows us to solve the portfolio optimization problem by maximizing over $\widetilde{\boldsymbol{\varpi}}_t$. Notice the wealth process (5.7) does not explicitly depend on X_t. The investor solves the following utility maximization problem

$$u_i(t,w) = \sup_{\widetilde{\boldsymbol{\varpi}}} \mathbb{E}^{\mathbb{P}}\left[U(W_{\widetilde{T}}^{\boldsymbol{\varpi}}) | W_t = w, \xi_t = i\right], \quad (5.8)$$

where

$$U(w) = -e^{-\gamma w}, \quad w \in \mathbb{R},$$

is exponential utility function with a constant risk aversion parameter $\gamma > 0$.

The investor's value function $u_i(t,w)$ is determined from a system of HJB equations. Precisely, we have

$$\partial_t u_i + \max_{\widetilde{\boldsymbol{\varpi}}_t}\left\{\left(\zeta_i\widetilde{\varpi}_t^{(0)} - \sum_{j\in E\setminus\{i\}} \widetilde{q}_{ij}\widetilde{\varpi}_t^{(j)}\right)\partial_w u_i + \frac{(\widetilde{\varpi}_t^{(0)})^2}{2}\partial_{ww}u_i\right.$$
$$\left. + \sum_{j\in E\setminus\{i\}} q_{ij}\left(u_j(t, w+\widetilde{\varpi}_t^{(j)}) - u_i(t,w)\right)\right\} = 0, \quad (5.9)$$

for $i \in E$ and $t \in [0,\widetilde{T})$. The terminal condition is $u_i(\widetilde{T},w) = -e^{-\gamma w}$, for $i \in E$. We have used the shorthand notations: $\widetilde{q}_{ij} \equiv \widetilde{q}(i,j)$, $q_{ij} \equiv q(i,j)$ and $\zeta_i \equiv \zeta(i)$.

Performing the optimization in the HJB equation (5.9) and assuming that $\partial_{ww}u_i \leq 0$ (which will be verified later), we obtain the first-order conditions for the optimal strategy:

$$\begin{cases} \widetilde{\varpi}^{(0)*}(t,w,i) = -\zeta_i \dfrac{\partial_w u_i(t,w)}{\partial_{ww}u_i(t,w)}, \\ \partial_w u_j(t, w+\widetilde{\varpi}^{(j)*}(t,w,i)) = \dfrac{\widetilde{q}_{ij}}{q_{ij}}\partial_w u_i(t,w), \end{cases} \tag{5.10}$$

for $i \in E$ and $j \in E \setminus \{i\}$. Plugging this into (5.9), the HJB equations become

$$\partial_t u_i - \frac{(\zeta_i \partial_w u_i)^2}{2\partial_{ww}u_i} - \sum_{j\in E\setminus\{i\}} \widetilde{q}_{ij}\widetilde{\varpi}_i^{(j)*}\partial_w u_i + \sum_{j\in E\setminus\{i\}} q_{ij}\big(u_j(t, w+\widetilde{\varpi}_i^{(j)*}) - u_i(t,w)\big) = 0, \tag{5.11}$$

for $i \in E$, where we have denoted $\widetilde{\varpi}_i^{(j)*} \equiv \widetilde{\varpi}^{(j)*}(t,w,i)$.

We now consider the transformation for the value function

$$u_i(t,w) = -e^{-\gamma w + \varphi_i(t)}. \tag{5.12}$$

Applying this to the first-order conditions (5.10) and HJB equation (5.11), the optimal strategy can be written explicitly

$$\begin{cases} \widetilde{\varpi}_i^{(0)*} = \dfrac{\zeta_i}{\gamma}, \\ \widetilde{\varpi}_i^{(j)*}(t) = -\dfrac{1}{\gamma}\left(\log\dfrac{\widetilde{q}_{ij}}{q_{ij}} + \varphi_i(t) - \varphi_j(t)\right), \end{cases} \tag{5.13}$$

for $i \in E$ and $j \in E \setminus \{i\}$. We note that both $\widetilde{\varpi}_i^{(0)*}$ and $\widetilde{\varpi}_i^{(j)*}(t)$ do not depend on wealth w.

Substituting the transformation (5.12) and optimal strategy (5.13) into (5.11), we obtain a system of ODEs for $\varphi_i(t)$:

$$\varphi_i^{\top}(t) - \sum_{j\in E\setminus\{i\}} \widetilde{q}_{ij}(\varphi_i(t) - \varphi_j(t)) - \alpha_i = 0, \tag{5.14}$$

where

$$\alpha_i = \frac{\zeta_i^2}{2} + \sum_{j\in E\setminus\{i\}} \widetilde{q}_{ij}\log\frac{\widetilde{q}_{ij}}{q_{ij}} - \widetilde{q}_{ij} + q_{ij}, \tag{5.15}$$

for $i \in E$ and $t \in [0, \widetilde{T})$. The terminal condition is

$$\varphi_i(\widetilde{T}) = 0, \quad \text{for } i \in E.$$

This ODE system admits the solution

$$\boldsymbol{\varphi}(t) = -\int_t^{\widetilde{T}} \exp(\widetilde{\boldsymbol{Q}}(\widetilde{T} - s))\boldsymbol{\alpha}\, ds,$$

where $\widetilde{\boldsymbol{Q}}$ is the generator matrix for ξ_t under measure $\mathbb{Q}$, and

$$\boldsymbol{\varphi}(t) = (\varphi_1(t), \ldots, \varphi_M(t))^\top,$$
$$\boldsymbol{\alpha} = (\alpha_1(t), \ldots, \alpha_M(t))^\top.$$

With this solution, along with transformation (5.12), direction calculations now show that the second-order condition $\partial_{ww}u_i = \gamma^2 u_i \leq 0$ is satisfied. Therefore, the solution to the HJB system (5.9) is indeed given by (5.12). In addition, the optimal strategy $\boldsymbol{\varpi}^*$ can be recovered from (5.6).

The ODE system (5.14) implies a probabilistic representation for $\varphi_i(t)$, given by

$$\varphi_i(t) = \mathbb{E}^{\mathbb{Q}}\left[-\int_t^{\widetilde{T}} \alpha(\xi_s)ds \,\middle|\, \xi_t = i\right], \tag{5.16}$$

where $\alpha(i) = \alpha_i$ is given by (5.15). Given that $x\log x - x + 1 \geq 0$, $\forall x > 0$, it follows that $\alpha(i) \geq 0$ and $\varphi_i(t) \leq 0$. Intuitively, $\exp(\varphi_i(t))$ acts like a discounting factor that depends directly on the regime-switching market price of risk. Since the exponential utility is negative and $\varphi_i(t) \leq 0$, this implies that $u_i(t, w) \geq -e^{-\gamma w}$, which means that the investor achieves a higher expected utility by dynamically trading the futures, as compared to only holding the same constant cash amount w.

Next, by putting the optimal strategy $\widetilde{\varpi}_i^{(0)*}$ and $\widetilde{\varpi}_i^{(j)*}(t)$ in (5.7), we obtain the optimal wealth process

$$\begin{aligned} dW_t^* = {} & \frac{1}{\gamma}\Bigg(\zeta^2(\xi_t) - \sum_{j\in E\setminus\{\xi_t\}} (q(\xi_t, j) - \widetilde{q}(\xi_t, j)) \\ & \times \left(\log\frac{\widetilde{q}(\xi_t, j)}{q(\xi_t, j)} + \varphi(t, \xi_t) - \varphi(t, j)\right)\Bigg)dt + \frac{\zeta(\xi_t)}{\gamma}dZ_t^{\mathbb{P}} \\ & - \frac{1}{\gamma}\int_{\mathbb{R}} \sum_{j\in E\setminus\{\xi_t\}} \left(\log\frac{\widetilde{q}(\xi_t, j)}{q(\xi_t, j)} + \varphi(t, \xi_t) - \varphi(t, j)\right) \\ & \times I_{\{z\in\Delta(\xi_t, j)\}} M^{\mathbb{P}}(dt, dz), \end{aligned} \tag{5.17}$$

where we have denoted $\varphi(t,i) = \varphi_i(t)$, for $i \in E$. The coefficients of the optimal wealth process in (5.17) do not depend on the spot price X_t but are modulated by the continuous-time Markov chain ξ_t. The optimal wealth is connected with the risk-neutral pricing measure and futures prices through the market price of risk $\zeta(\xi_t)$. The investor's risk aversion parameter γ also affects the optimal wealth. In particular, a higher γ will reduce the magnitude of the drift and volatility of the optimal wealth process.

Remark 5.1. When $M = 2$, the ODE system (5.14) admits an explicit solution

$$\varphi_i(t) = -\frac{1}{\widetilde{\lambda}_1+\widetilde{\lambda}_2}\left((\widetilde{\lambda}_2\alpha_1+\widetilde{\lambda}_1\alpha_2)(\widetilde{T}-t)+\widetilde{\lambda}_i(\alpha_i-\alpha_j)\frac{1-e^{-(\widetilde{\lambda}_1+\widetilde{\lambda}_2)(\widetilde{T}-t)}}{\widetilde{\lambda}_1+\widetilde{\lambda}_2}\right), \tag{5.18}$$

where $\widetilde{\lambda}_i = \widetilde{q}_{ij}$, for $(i,j) \in \{(1,2),(2,1)\}$ and $t \in [0,\widetilde{T}]$. In turn, the optimal strategy is given by

$$\begin{bmatrix} \varpi_i^{(1)*}(t,x) \\ \varpi_i^{(2)*}(t,x) \end{bmatrix} = -\frac{1}{\begin{aligned}&\gamma b(t,x,i)((F_j^{(2)}(t,x)-F_i^{(2)}(t,x))\partial_x F_i^{(1)}(t,x)\\&-(F_j^{(1)}(t,x)-F_i^{(1)}(t,x))\partial_x F_i^{(2)}(t,x))\end{aligned}} \times \begin{bmatrix} -\zeta_i(F_j^{(2)}(t,x)-F_i^{(2)}(t,x)) \\ \quad -b(t,x,i)\partial_x F_j^{(2)}(t,x)(\log\frac{\widetilde{\lambda}_i}{\lambda_i}+\varphi_i(t)-\varphi_j(t)) \\ \zeta_i(F_j^{(1)}(t,x)-F_i^{(1)}(t,x)) \\ \quad +b(t,x,i)\partial_x F_j^{(1)}(t,x)(\log\frac{\widetilde{\lambda}_i}{\lambda_i}+\varphi_i(t)-\varphi_j(t)) \end{bmatrix}, \tag{5.19}$$

where $\lambda_i = q_{ij}$, for $(i,j) \in \{(1,2),(2,1)\}$ and $(t,x) \in [0,\widetilde{T}] \times \mathbb{R}$.

5.3.2 *Certainty equivalent*

In order to quantify the value of trading futures to the investor, we define the investor's certainty equivalent associated with the utility maximization problem. The certainty equivalent is the guaranteed cash amount that would yield the same utility as that from dynamically trading futures according to (5.8). This amounts to applying the inverse of the utility function to the value function in (5.12), that is,

$$CE_i(t, w) := U^{-1}(u_i(t, w)) = w - \frac{\varphi_i(t)}{\gamma}. \tag{5.20}$$

Therefore, the certainty equivalent is the sum of the investor's wealth w and a time-dependent component $-\frac{\varphi_i(t)}{\gamma}$. From the probabilistic representation in (5.16), we know that $\varphi_i(t)$ is negative. The certainty equivalent is also inversely proportional to the risk aversion parameter γ, which means that a more risk averse investor has a lower certainty equivalent, valuing the futures trading opportunity less.

Under our regime-switching framework, the certainty equivalent is the same across different underlying models as long as the market prices of risk stay the same. Therefore, without picking a specific model, we can still compute the certainty equivalent. Figure 5.1 illustrates the certainty equivalents under two risk aversion levels under two regimes in a two-regime market. Under each regime, the less risk averse investor ($\gamma = 0.5$) has a higher certainty equivalent than the more risk averse investor ($\gamma = 2$).

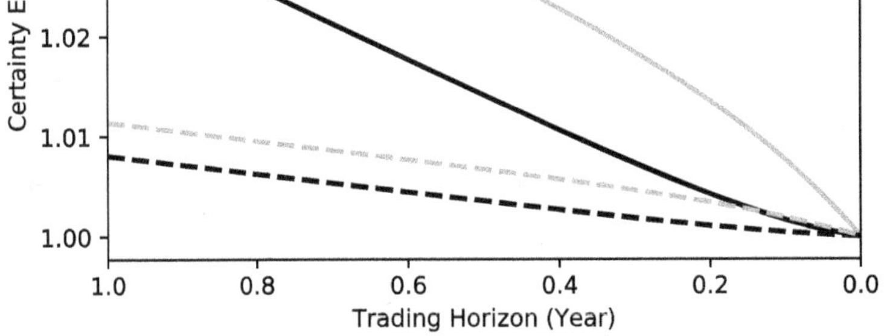

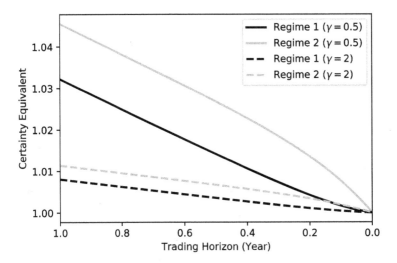

Fig. 5.1. The certainty equivalents corresponding to two different risk aversion levels in a two-regime market. They are plotted as functions of the trading horizon, for $\gamma = 0.5$ (solid lines) and $\gamma = 2$ (dashed lines) in regime 1 (dark color) and regime 2 (light color). Common parameters are $\widetilde{q}_{12} = q_{12} = 0.8$, $\widetilde{q}_{21} = q_{21} = 0.6$, $w = 1$, $\zeta_1 = 0.1$ and $\zeta_2 = 0.3$.

According to the parameters for the two regimes, Regime 2 has a higher market price of risk than regime 1, which explains that the investor's certainty equivalent is higher in regime 2 than in regime 1. All else being equal, the certainty equivalent is higher when there is more time to trade. Hence, as the trading horizon reduces to zero, the certainty equivalent converges to the initial wealth w, which is set to be 1 in this example. This means that the second term in (5.20) converges to zero since $\varphi_i(t) \to 0$ as $t \to \widetilde{T}$.

In the next two sections, we consider two model specifications and provide the corresponding numerical examples.

5.4 Regime-Switching Geometric Brownian Motion

Suppose the log-price of the underlying asset follows the SDE:

$$dX_t = \mu(\xi_t)dt + \sigma(\xi_t)dZ_t^{\mathbb{Q}},$$

under the risk-neutral measure $\mathbb{Q}$. We call this model the RS-GBM because without ξ the spot price S is simply a GBM. This model belongs to our regime-switching framework discussed in the previous section. Indeed, this amounts to setting the coefficients in SDE (4.1) to be

$$\widetilde{a}(t, X_t, \xi_t) = \mu(\xi_t), \quad \text{and} \quad b(t, X_t, \xi_t) = \sigma(\xi_t). \tag{5.21}$$

Substituting (5.21) into (4.2), we obtain the PDE system for the futures price function under this model. Precisely, for $i \in E$, we have

$$\partial_t F_i + \mu_i \partial_x F_i + \frac{\sigma_i^2}{2}\partial_{xx} F_i + \sum_{j \in E\setminus\{i\}} \widetilde{q}(i,j)(F_j - F_i) = 0,$$

with $\mu_i = \mu(i)$ and $\sigma_i = \sigma(i)$. The terminal condition is $F_i(T, x) = e^x$, $x \in \mathbb{R}$.

Under this model, the futures price admits the separation of variables:

$$F_i(t, x) = e^x\, g_i(t),$$

where $(g_i(t))_{i=1,\dots,M}$ solve the system of ODEs:

$$\frac{dg_i(t)}{dt} + (\mu_i + \frac{\sigma_i^2}{2})g_i(t) + \sum_{j \in E\setminus\{i\}} \widetilde{q}(i,j)(g_j(t) - g_i(t)) = 0,$$

for $t \in [0, T)$, with the terminal condition $g_i(T) = 1$, for $i = 1, \ldots, M$. Defining $\boldsymbol{g}(t) = (g_1(t), \ldots, g_M(t))^\top$, we can write the solution as

$$\boldsymbol{g}(t) = \exp((\boldsymbol{G} + \widetilde{\boldsymbol{Q}})(T - t))\mathbf{1},$$

where $\widetilde{\boldsymbol{Q}}$ is the generator matrix under the measure $\mathbb{Q}$ and

$$\boldsymbol{G} = \operatorname{diag}\left(\frac{2\mu_1 + \sigma_1^2}{2}, \frac{2\mu_2 + \sigma_2^2}{2}, \ldots, \frac{2\mu_M + \sigma_M^2}{2}\right).$$

In addition, the ODE system (4.7) implies a probabilistic representation for $g_i(t)$:

$$g_i(t) = \mathbb{E}^{\mathbb{Q}}\left[\exp\left(\int_t^T \mu(\xi_s) + \frac{\sigma^2(\xi_s)}{2} ds\right) \middle| \xi_t = i\right].$$

To sum up, the futures price in regime i is given by

$$F_i(t, x) = \exp(x)(\exp((\boldsymbol{G} + \widetilde{\boldsymbol{Q}})(T - t))\mathbf{1})_i,$$

where the subscript i denotes the ith entry of the vector.

In turn, using (4.4), the futures price satisfies the SDE

$$\begin{aligned} dF_t = &\Bigg(\sigma(\xi_t)F(t, X_t, \xi_t)\zeta(\xi_t) \\ &+ \sum_{j \in E \setminus \{\xi_t\}} (q(\xi_t, j) - \widetilde{q}(\xi_t, j))(F(t, X_t, j) - F(t, X_t, \xi_t))\Bigg)dt \\ &+ \sigma(\xi_t)F(t, X_t, \xi_t)dZ_t^{\mathbb{P}} \\ &+ \int_{\mathbb{R}} \sum_{j \in E \setminus \{\xi_t\}} (F(t, X_t, j) - F(t, X_t, \xi_t))I_{\{z \in \Delta(\xi_t, j)\}} M^{\mathbb{P}}(dt, dz). \end{aligned}$$

Next, we consider a portfolio of futures with M different maturities $T_1 < T_2 < \cdots < T_M$. If the associated coefficient matrix $\boldsymbol{\Gamma}$ is invertible, we can apply results in the Section 5.3. To that end, we have following proposition for coefficient matrix $\boldsymbol{\Gamma}$.

Proposition 5.2. *If* $\boldsymbol{\Gamma}(t_0, x, i)$ *is invertible for some specific time* $t_0 \leq T_1$, *then* $\boldsymbol{\Gamma}(t, x, i)$ *is invertible for any time* $t \leq T_1$.

Proof. By (5.2), the coefficient matrix is given by

$$\mathbf{\Gamma}(t,x,i) = e^x \begin{bmatrix} \sigma_i g_i^{(1)}(t) & \sigma_i g_i^{(2)}(t) & \cdots & \sigma_i g_i^{(M)}(t) \\ g_1^{(1)}(t) - g_i^{(1)}(t) & g_1^{(2)}(t) - g_i^{(2)}(t) & \cdots & g_1^{(M)}(t) - g_i^{(M)}(t) \\ \vdots & \vdots & \vdots & \vdots \\ g_{i-1}^{(1)}(t) - g_i^{(1)}(t) & g_{i-1}^{(2)}(t) - g_i^{(2)}(t) & \cdots & g_{i-1}^{(M)}(t) - g_i^{(M)}(t) \\ g_{i+1}^{(1)}(t) - g_i^{(1)}(t) & g_{i+1}^{(2)}(t) - g_i^{(2)}(t) & \cdots & g_{i+1}^{(M)}(t) - g_i^{(M)}(t) \\ \vdots & \vdots & \ddots & \vdots \\ g_M^{(1)}(t) - g_i^{(1)}(t) & g_M^{(2)}(t) - g_i^{(2)}(t) & \cdots & g_M^{(M)}(t) - g_i^{(M)}(t) \end{bmatrix},$$

for $i \in E$. The determinant of coefficient matrix $\mathbf{\Gamma}$ is denoted by

$$\Phi_i(t,x) = \det \mathbf{\Gamma}(t,x,i), \quad i \in E.$$

We define the matrix $\widetilde{\mathbf{\Gamma}}$ by adding $1/\sigma_i$ of the first row to other rows in matrix $\mathbf{\Gamma}$. Then, we define the matrix $\boldsymbol{H_k}$ by taking t-derivative in the kth row of matrix $\widetilde{\mathbf{\Gamma}}$. For example,

$$\boldsymbol{H_1}(t,x,i) = e^x \begin{bmatrix} \sigma_i \frac{d}{dt} g_i^{(1)}(t) & \sigma_i \frac{d}{dt} g_i^{(2)}(t) & \cdots & \sigma_i \frac{d}{dt} g_i^{(M)}(t) \\ g_1^{(1)}(t) & g_1^{(2)}(t) & \cdots & g_1^{(M)}(t) \\ \vdots & \vdots & \vdots & \vdots \\ g_{i-1}^{(1)}(t) & g_{i-1}^{(2)}(t) & \cdots & g_{i-1}^{(M)}(t) \\ g_{i+1}^{(1)}(t) & g_{i+1}^{(2)}(t) & \cdots & g_{i+1}^{(M)}(t) \\ \vdots & \vdots & \ddots & \vdots \\ g_M^{(1)}(t) & g_M^{(2)}(t) & \cdots & g_M^{(M)}(t) \end{bmatrix}.$$

Then, holding the log-price x fixed, we differentiate to get

$$\frac{d}{dt}\Phi_i(t,x) = \frac{d}{dt}\det \widetilde{\mathbf{\Gamma}}(t,x,i) = \sum_{k=1}^{M} \det \boldsymbol{H_k}(t,x,i). \tag{5.22}$$

Applying (4.7) and

$$\sum_{j \in E \setminus \{i\}} \widetilde{q}(i,j) = -\widetilde{q}(i,i),$$

we get

$$\sum_{k=1}^{M} \det \boldsymbol{H_k}(t,x,i) = \sum_{k=1}^{M} -\left(\mu_k + \frac{\sigma_k^2}{2} + \widetilde{q}(k,k)\right)\Phi_i(t,x), \tag{5.23}$$

for $i \in E$.

Combining (5.22) and (5.23), we have

$$\frac{d}{dt}\Phi_i(t,x) + \sum_{k=1}^{M}\left(\mu_k + \frac{\sigma_k^2}{2} + \widetilde{q}(k,k)\right)\Phi_i(t,x) = 0.$$

Then, for any $t_0, t \le T_1$, $\Phi_i(t,x)$ satisfies

$$\Phi_i(t,x) = \exp\left(-\sum_{k=1}^{M}\left(\mu_k + \frac{\sigma_k^2}{2} + \widetilde{q}(k,k)\right)(t-t_0)\right)\Phi_i(t_0,x). \tag{5.24}$$

Thus, if $\boldsymbol{\Gamma}(t_0,x,i)$ is invertible for some specific time $t_0 \le T_1$, then $\boldsymbol{\Gamma}(t,x,i)$ is invertible for any time $t \le T_1$. □

Example 5.3 (Market with Two Regimes). Assume a market with two regimes, i.e. $E = \{1,2\}$. The coefficient matrix $\boldsymbol{\Gamma}$ for futures pair $(F^{(1)}, F^{(2)})$ is given explicitly by

$$\boldsymbol{\Gamma}(t,x,i) = e^x \begin{bmatrix} \sigma_i g_i^{(1)}(t) & \sigma_i g_i^{(2)}(t) \\ g_j^{(1)}(t) - g_i^{(1)}(t) & g_j^{(2)}(t) - g_i^{(2)}(t) \end{bmatrix},$$

for $(i,j) \in \{(1,2),(2,1)\}$. Then, the matrix determinant $\Phi_i(t,x)$ is also explicit:

$$\Phi_i(t,x) = e^{2x}\sigma_i(g_i^{(1)}(t)g_j^{(2)}(t) - g_j^{(1)}(t)g_i^{(2)}(t)),$$

for $(i,j) \in \{(1,2),(2,1)\}$.

Applying Proposition 5.2, the coefficient matrix $\boldsymbol{\Gamma}$ is invertible for any time $t \le T_2$, if and only if

$$\Phi_i(T_2,x) = e^{2x}\sigma_i(g_j^{(2)}(T_2) - g_i^{(2)}(T_2)) \ne 0,$$

which is equivalent to

$$\mu_1 + \frac{\sigma_1^2}{2} \ne \mu_2 + \frac{\sigma_2^2}{2},$$

according to the probabilistic representation (4.8).

If $\mu_1 + \sigma_1^2/2 \neq \mu_2 + \sigma_2^2/2$, then we can apply the results in Section 5.3. The value function $u_i(t, w)$ satisfies

$$u_i(t, w) = -e^{-\gamma w + \varphi_i(t)},$$

where $\varphi_i(t)$ is given by (5.18). The optimal strategy is described by the positions in the two futures in each state. An application of (5.19) yields the optimal strategy

$$\begin{bmatrix} \varpi_i^{(1)*}(t,x) \\ \varpi_i^{(2)*}(t,x) \end{bmatrix} = -\frac{e^{-x}}{\gamma\sigma_i(g_i^{(1)}(t)g_j^{(2)}(t) - g_j^{(1)}(t)g_i^{(2)}(t))}$$
$$\times \begin{bmatrix} -\zeta_i(g_j^{(2)}(t) - g_i^{(2)}(t)) - \sigma_i g_i^{(2)}(t)\left(\log\frac{\widetilde{\lambda}_i}{\lambda_i} + \varphi_i(t) - \varphi_j(t)\right) \\ \zeta_i(g_j^{(1)}(t) - g_i^{(1)}(t)) + \sigma_i g_i^{(1)}(t)\left(\log\frac{\widetilde{\lambda}_i}{\lambda_i} + \varphi_i(t) - \varphi_j(t)\right) \end{bmatrix},$$

for $(i, j) \in \{(1, 2), (2, 1)\}$ and $(t, x) \in [0, \widetilde{T}] \times \mathbb{R}$.

If the parameters values happen to satisfy

$$\mu_1 + \frac{\sigma_1^2}{2} = \mu_2 + \frac{\sigma_2^2}{2},$$

then the futures prices will be the same in two states according to (4.8). In this special case, the futures price SDE becomes

$$dF_t = \sigma(\xi_t)F(t, X_t, \xi_t)\zeta(\xi_t)dt + \sigma(\xi_t)F(t, X_t, \xi_t)dZ_t^{\mathbb{P}}.$$

It is important to note that, in contrast to (4.10), the futures price process has continuous paths even though the price process is modulated by a Markov chain.

In turn, the associated futures portfolio optimization can be solved as well. Interestingly the cash amount of the optimal futures position, that is,

$$\varpi_i^{(1)*}(t,x)F_i^{(1)}(t,x) + \varpi_i^{(2)*}(t,x)F_i^{(2)}(t,x) = \zeta(i)/\gamma\sigma_i$$

does not depend on t or x. This also means that the value of the position stays constant during each regime.

To illustrate the futures trading problem under this model, we simulate the sample paths for spot price, futures price, optimal investment, and optimal wealth process. The regime switching between two states, with transition probabilities $q_{12} = 2$ and $q_{21} = 4$ which are entries of generator matrix $\boldsymbol{Q}$. The trading horizon $\widetilde{T} = 0.6$ which is no greater than the maturity of the futures contract $T_1 = 0.6$ and $T_2 = 0.8$. All parameters are summarized in the Table 4.1.

In Figure 5.2, we illustrate the sample paths of the optimal positions in the two futures over the trading horizon. As we can see, the optimal positions are of opposite signs, meaning that the investor will go long on the T_1-futures and short on the T_2-futures. Long-short strategies are very common in futures trading, and they can help mitigate the effect of regime switching. As the market switches from regime 2 to regime 1 at time t_1, the magnitude of the futures position is reduced immediately. The investor then take larger long-short positions when the market returns to regime 2 from regime 1 at time t_2. According to the sample paths, the optimal positions tend to decay in time during each regime, meaning that the investor gradually reduces investments towards the end of the trading horizon.

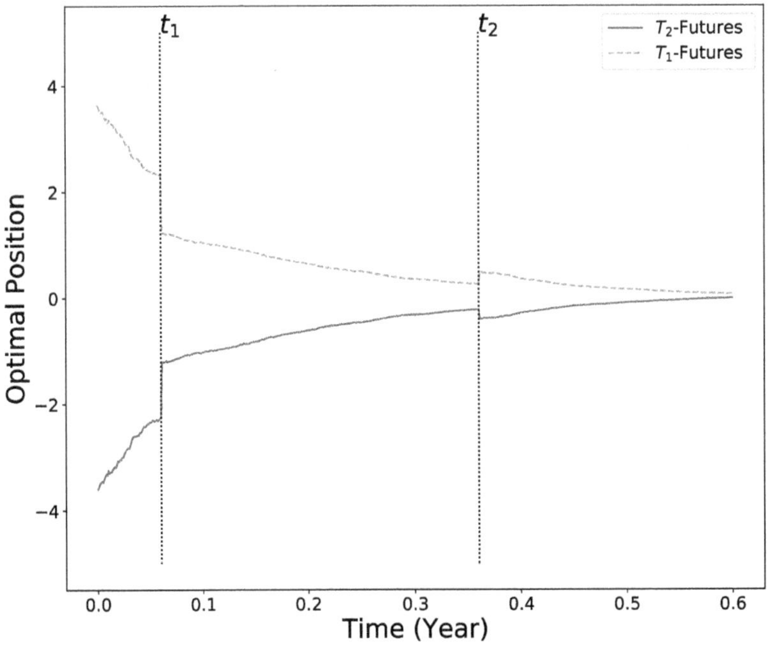

Fig. 5.2. Sample paths of the optimal positions $(\varpi^{(1)}, \varpi^{(2)})$ in the T_1-futures and T_2-futures respectively under the RS-GBM model. The futures positions tend to decrease over time and approaches zero near the end of the trading horizon. Jumps in both positions occur at the regiem-switching times t_1 and t_2.

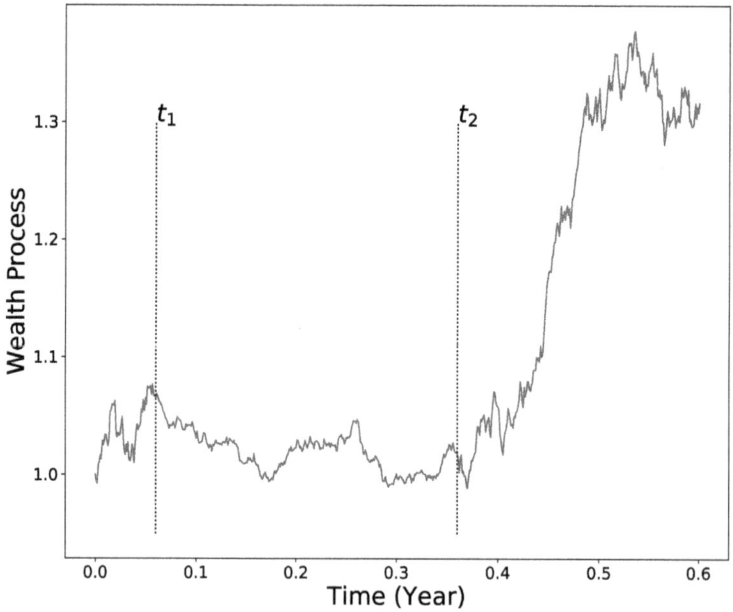

Fig. 5.3. Sample path of the optimal wealth process over the trading horizon under the RS-GBM model.

Figure 5.3 shows the wealth process corresponding to the optimal trading strategy in Figure 5.2. The long-short strategy tends to reduce the shock in portfolio value due to regime switching. Therefore, no sizable jumps in the wealth process are observed at the regime-switching times t_1 and t_2. As we can see, the wealth process decreases slightly in regime 1. This is intuitive since regime 1 has lower risk premium than in regime 2. The wealth increases sharply in regime 2 towards the end of the trading horizon, even though the positions are gradually reduced over time.

In Figure 5.4, we plot the sample path for the absolute value of coefficient matrix determinant $|\Phi|$ corresponding to the sample paths of the spot and futures prices. Recall from (5.24), $|\Phi|$ is exponentially increasing or decreasing overtime. Moreover, in our numerical settings, the term $\widetilde{q}(i,i) = -\widetilde{q}(i,j)$ dominates the term $\mu_i + \sigma_i^2/2$. Therefore, $|\Phi|$ appears to be exponentially increasing in time. Since $|\Phi|$ is not equal to 0, Γ is invertible and we are able to apply the results in Section 5.3 for the RS-GBM model.

In Figure 5.5, we randomly generate 100 sample paths for $|\Phi|$, the absolute value of coefficient matrix determinant. Recall that in equation (5.24),

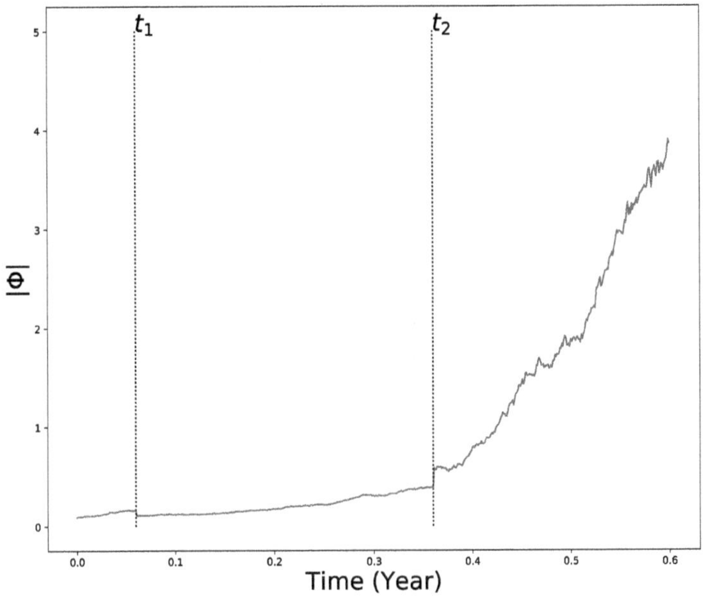

Fig. 5.4. Sample path of the absolute value of the determinant Φ for the coefficient matrix Γ under the RS-GBM model.

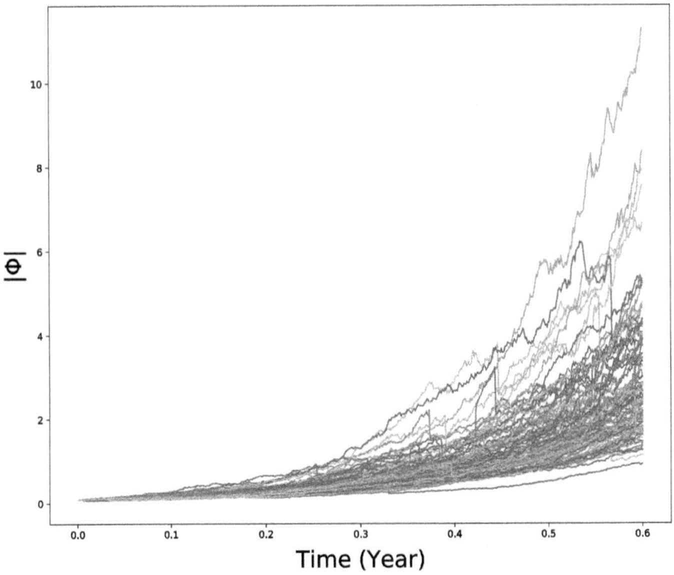

Fig. 5.5. 100 sample paths of absolute value of coefficient matrix determinant $|\Phi|$.

for fixed x, $|\Phi|$ is exponentially increasing or decreasing overtime. Moreover, in our numerical illustration, the term $\widetilde{q}(i,i) = -\widetilde{q}(i,j)$ dominates the term $\mu_i + \sigma_i^2/2$. Thus, $|\Phi|$ appears exponentially increasing with respect to time. In addition, we are able to apply the results in Section 5.3.

5.5 Regime-Switching Exponential Ornstein–Uhlenbeck Model

As is well known, the exponential Ornstein–Uhlenbeck (RS-XOU) process and its variations are widely used to model commodity prices. As an generalization of the RS-XOU, we now consider the regime-switching RS-XOU model and illustrate the optimal trading strategies under this model. In the RS-XOU model, the log spot price evolves according to

$$dX_t = \kappa(\xi_t)(\theta(\xi_t) - X_t)dt + \sigma(\xi_t)dZ_t^{\mathbb{Q}}, \tag{5.25}$$

where $\kappa(\xi_t)$, $\theta(\xi_t)$ and $\sigma(\xi_t)$ are the functions of regimes. This amounts to simply setting

$$\widetilde{a}(t, X_t, \xi_t) = \kappa(\xi_t)(\theta(\xi_t) - X_t)$$

and

$$b(t, X_t, \xi_t) = \sigma(\xi_t)$$

in (4.1). Therefore, the results under the general regime-switching model can be directly applied to the RS-XOU model.

5.5.1 *Futures dynamics and utility maximization*

The futures price function $F_i(t,x)$ satisfies PDE (4.2). Substituting (5.25) into (4.4), we obtain the $\mathbb{Q}$-dynamics for futures price F_t:

$$\begin{aligned} dF_t &= \eta(t, X_t, \xi_t)dZ_t^{\mathbb{Q}} \\ &\quad + \int_{\mathbb{R}} \sum_{j \in E \setminus \{\xi_t\}} (F(t, X_t, j) - F(t, X_t, \xi_t)) I_{\{z \in \Delta(\xi_t, j)\}} M^{\mathbb{Q}}(dt, dz), \end{aligned}$$

where the volatility term is given by

$$\eta(t, x, i) = \sigma_i \partial_x F_i(t, x).$$

In addition, under the measure $\mathbb{P}$,

$$
\begin{aligned}
dF_t = \Bigg(&\zeta(\xi_t)\eta(t, X_t, \xi_t) \\
&+ \sum_{j\in E\setminus\{\xi_t\}} (q(\xi_t, j) - \widetilde{q}(\xi_t, j))(F(t, X_t, j) - F(t, X_t, \xi_t))\Bigg) dt \\
&+ \eta(t, X_t, \xi_t) dZ_t^{\mathbb{P}} + \int_{\mathbb{R}} \sum_{j\in E\setminus\{\xi_t\}} (F(t, X_t, j) - F(t, X_t, \xi_t)) \\
&\times I_{\{z\in\Delta(\xi_t, j)\}} M^{\mathbb{P}}(dt, dz).
\end{aligned}
$$

Example 5.4 (Market with Two Regimes). We consider a portfolio of futures with two different maturities T_1 and T_2 in a two-state market, i.e. $E = \{1, 2\}$. The associated coefficient matrix is given by

$$
\mathbf{\Gamma}(t, x, i) = \begin{bmatrix} \sigma_i \partial_x F_i^{(1)}(t, x) & \sigma_i \partial_x F_i^{(2)}(t, x) \\ F_j^{(1)}(t, x) - F_i^{(1)}(t, x) & F_j^{(2)}(t, x) - F_i^{(2)}(t, x) \end{bmatrix},
$$

for $(i, j) \in \{(1, 2), (2, 1)\}$. Moreover, we assume $\mathbf{\Gamma}(t, x, i)$ be invertible. Under this assumption, we can apply the result in Section 5.3. The value function

$$
u_i(t, w) = -e^{-\gamma w + \varphi_i(t)},
$$

where $\varphi_i(t)$ is given by (5.18), holds.

Applying (5.19), we immediately obtain the optimal strategy

$$
\begin{aligned}
&\begin{bmatrix} \varpi_i^{(1)*}(t, x) \\ \varpi_i^{(2)*}(t, x) \end{bmatrix} \\
&= -\frac{1}{\begin{aligned}\gamma\sigma_i((F_j^{(2)}(t, x) - F_i^{(2)}(t, x))\partial_x F_i^{(1)}(t, x) \\ - (F_j^{(1)}(t, x) - F_i^{(1)}(t, x))\partial_x F_i^{(2)}(t, x))\end{aligned}} \\
&\quad \times \begin{bmatrix} -\zeta_i(F_j^{(2)}(t, x) - F_i^{(2)}(t, x)) - \sigma_i \partial_x F_j^{(2)}(t, x)(\log \frac{\widetilde{\lambda}_i}{\lambda_i} + \varphi_i(t) - \varphi_j(t)) \\ \zeta_i(F_j^{(1)}(t, x) - F_i^{(1)}(t, x)) + \sigma_i \partial_x F_j^{(1)}(t, x)(\log \frac{\widetilde{\lambda}_i}{\lambda_i} + \varphi_i(t) - \varphi_j(t)) \end{bmatrix},
\end{aligned}
$$

for $(i, j) \in \{(1, 2), (2, 1)\}$ and $(t, x) \in [0, \widetilde{T}] \times \mathbb{R}$.

Remark 5.5. If the speed of mean reversion is identical in all regimes, i.e. $\kappa_i = \kappa\ \forall i$, then the futures price admits the following form

$$F_i(t,x) = \exp\left(e^{-\kappa(T-t)}x\right)h_i(t), \tag{5.26}$$

where $(h_i(t))_{i=1,\ldots,M}$ solves the ODE system

$$\begin{aligned}\frac{h_i(t)}{dt} &+ \left(\kappa\theta_i e^{-\kappa(T-t)} + \frac{\sigma_i^2}{2}e^{-2\kappa(T-t)}\right)h_i(t)\\ &+ \sum_{j\in E\setminus\{i\}} \widetilde{q}(i,j)(h_j(t) - h_i(t)) = 0,\end{aligned}$$

with the terminal condition

$$h_i(T) = 1, \quad \text{for } i = 1,\ldots,M.$$

The price formula (5.26) means that

$$\eta(t,x,i) = \sigma_i e^{-\kappa(T-t)}F_i(t,x)$$

in (4.4). As a result, the futures price satisfies the SDE

$$\begin{aligned}dF_t &= \sigma(\xi_t)e^{-\kappa(T-t)}\exp(e^{-\kappa(T-t)}X_t)h(t,\xi_t)dZ_t^{\mathbb{Q}}\\ &\quad + \exp(e^{-\kappa(T-t)}X_t)\int_{\mathbb{R}}\sum_{j\in E\setminus\{\xi_t\}}(h(t,j) - h(t,\xi_t))\\ &\quad \times I_{\{z\in\Delta(\xi_t,j)\}}M^{\mathbb{Q}}(dt,dz),\end{aligned}$$

where we have denoted $h(t,i) \equiv h_i(t)$.

5.5.2 *Numerical implementation and examples*

We now simulate the sample paths for spot price, futures prices, optimal investment, and optimal wealth process. The numerical procedure to compute the futures price under the RS-XOU model can be found in Chapter 4.3.

The regime switching between two states, with transition probabilities $q_{12} = 2$ and $q_{21} = 4$ which are entries of generator matrix $\boldsymbol{Q}$. The trading horizon $\widetilde{T} = 0.6$, and the two futures contracts have maturities $T_1 = 0.6$ and $T_2 = 0.8$. All parameters are summarized in Table 4.2. As shown in Figure 4.3 in the last chapter, the market starts in regime 2, then switches to regime 1 at time t_1 before returning to regime 2 at time t_2.

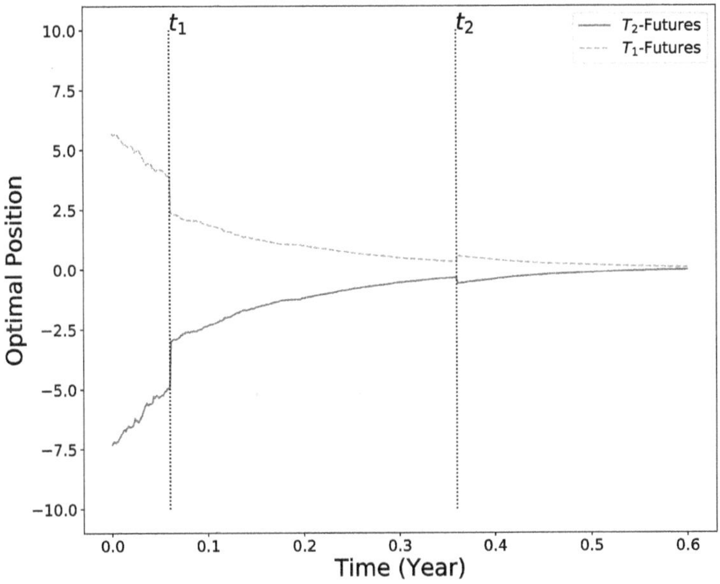

Fig. 5.6. Sample paths of the optimal positions $(\varpi^{(1)}, \varpi^{(2)})$ in the T_1-futures and T_2-futures respectively under the RS-XOU model. The futures positions tend to decrease over time and approaches zero near the end of the trading horizon. Jumps in both positions occur at the regime-switching times t_1 and t_2.

Figure 5.6 shows the sample path of the optimal positions in the two futures over the trading horizon. The optimal strategy is to long T_1-futures and short T_2-futures. The long-short positions help reduce the effect of regime switching. As the market switches from regime 2 to regime 1 at time t_1, the magnitude of the futures position is immediately reduced. The investor then take larger positions when the market returns to regime 2 from regime 1 at time t_2. According to the sample paths, the optimal positions tend to decay in time during each regime, meaning that the investor gradually reduces investments towards the end of the trading horizon.

Figure 5.7 shows the optimal wealth process over time. By placing opposite position in two futures, we reduce the shock in portfolio value due to regime switching. Therefore, we do not observe sizable jumps in the wealth process at the regime-switching times t_1 and t_2. Just like the futures and spot prices, the wealth process tends to decrease in regime 1 and increase in regime 2 (toward the end of the trading horizon).

Lastly in Figure 5.8, we plot the sample path for $|\Phi|$, which is the absolute value of coefficient matrix determinant Γ. The strict positivity of

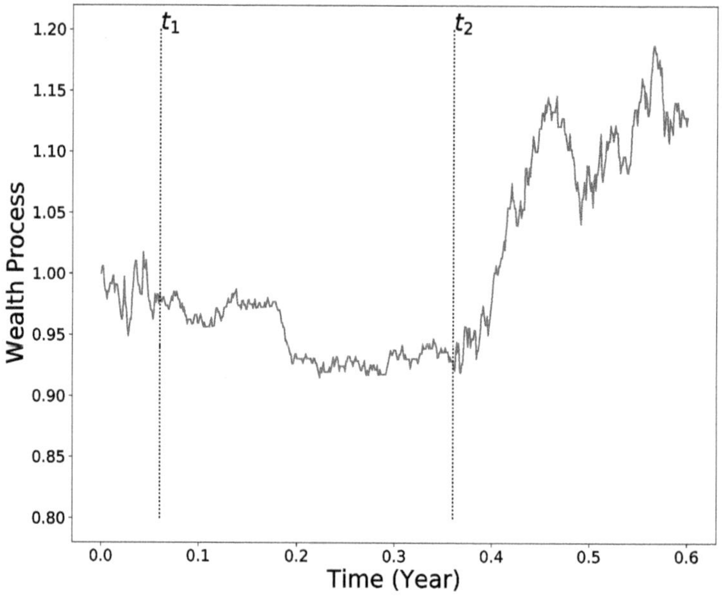

Fig. 5.7. Sample path of the optimal wealth process over the trading horizon under the RS-XOU model.

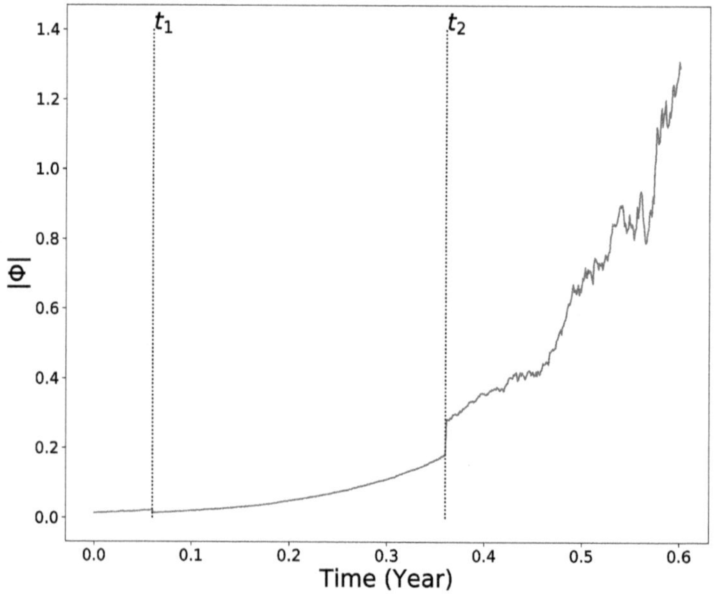

Fig. 5.8. Sample path of the absolute value of the determinant Φ for the coefficient matrix Γ under the RS-XOU model.

$|\Phi|$ informs us that Γ is invertible and thus the results in Section 5.3 can be applied. Like in Figure 5.4, $|\Phi|$ tends to increase exponentially in time.

5.6 Conclusion

We have analyzed the problem of dynamically trading a portfolio of futures in a regime-switching market. Under a general market framework, the portfolio optimization problem leads to the analytical and numerical studies of a system of HJB equations, which are solved explicitly. The optimal trading strategies and optimal wealth process are also given analytically and illustrated numerically. Our methodology has been applied to the RS-GBM and RS-XOU models, but it can also be used for other model specifications within the framework or adapted to study models with additional factors.

Moreover, one can also utilize our framework to examine other classes of trading strategies, such as rolling and timing strategies. We refer to Leung and Li (2016) and Leung *et al.* (2016) for such strategies under mean-reverting models without regime switching. Another major application of futures portfolios is for tracking or gaining exposure to a commodity or index (see, e.g. Leung and Ward (2018) and references therein). It would be of practical interest to examine the effect of regime switching on the strategies and tracking errors. Other than futures portfolio, it is also useful to study the dynamic trading of other derivatives, such as options and swaps, in a regime-switching market.

Chapter 6

Dynamic Futures Portfolios with Stochastic Basis

In the previous chapters, the futures portfolios consist of multiple futures contracts written on the same underlying asset. A portfolio that has exposures to multiple underlying assets within the same asset class or across different asset classes offers more degrees of freedom in portfolio construction and certain diversification effects. Nevertheless, additional underlying assets and futures increase the dimension and complexity of the portfolio optimization problem.

Motivated by these observations, we introduce in this chapter a novel way to directly model the joint price dynamics of the underlying assets and associated futures. It is of practical interest to develop a stochastic model that can capture the correlation among all the futures and underlying assets while maintaining analytical tractability, numerically efficiency, as well as interpretability.

For each futures contract, the spread between the two prices is called the *basis*. By no-arbitrage theory, futures prices must converge to the spot price at expiry, so the basis process is expected to converge to zero as the associated futures contract expires as well. This behavior of the basis process leads us to model the random basis directly using a Brownian bridge and in turn express each futures price process through the associated basis process. Working with multiple futures contracts, we introduce a *multi-dimensional* Brownian bridge model, where each component converges to zero at the respective maturity.

The random end point of a randomized Brownian bridge can be used to describe the non-convergence of futures. This is a common observation among several agricultural futures, where the futures prices may deviate from the spot price at expiry due to a number of factors, including storage costs. This non-convergence phenomenon has been illustrated in a few studies, including Irwin *et al.* (2011), Adjemian *et al.* (2013), and Garcia *et al.* (2015). We refer to Guo and Leung (2017) for a stochastic model for capturing this price behavior and pricing agricultural futures.

In Section 6.2, we analyze the problem of dynamically trading futures with different underlying assets under the stochastic basis model. Compared to the previous chapters, another new element of our utility maximization problem is the incorporation of portfolio constraints on the futures positions. Our general setup captures the dollar neutral and market neutral constraints, which are widely used in industry. The optimal strategies for both unconstrained and constrained cases are derived. This is achieved by solving the associated Hamilton–Jacobi–Bellman (HJB) equations. We also provide verification theorem that confirms the solution to value function is indeed the solution to the associated HJB equation. Moreover, we show that the original HJB equations are reduced to a system of linear ODEs and the investor's value function admits a separable form under our stochastic basis framework.

In Section 6.3, we define the certainty equivalents corresponding to portfolios with different constraints. This allows us to quantify the impact of portfolio constraints on the value of futures trading. In addition, we solve for the optimal constraint that maximizes the certainty equivalent for the risk-averse portfolio manager. Numerical examples are provided in Section 6.4 to illustrate how certainty equivalent depend on number of traded futures and different portfolio constraints.

Our stochastic basis model is closest to that introduced by Angoshtari and Leung (2019b, 2020). They analyze the problem of dynamically trading a futures contract and its underlying asset. The associated basis is modeled by a Brownian bridge, but the process is stopped early to capture the non-convergence of prices at the end of trading horizon. In comparison to these works, the portfolio optimization problem herein involves trading futures only and the underlying assets are not traded. This portfolio setup is more realistic as many underlying assets, ranging from commodities to indices, are not directly or liquidly traded. Another important difference is that the portfolios considered in this chapter can be subject to constraints that limit some or all of the futures positions. Also in a number of companion papers (Leung and Yan, 2018, 2019; Leung and Zhou, 2019), the utility maximization approach is used to derive dynamic futures trading strategies under various stochastic models without portfolio constraints.

6.1 Multidimensional Brownian Bridge Model

We fix a physical probability space $(\Omega, \mathcal{F}, \mathbb{P})$, where $\mathbb{P}$ is the physical probability measure. The market consists of M risky underlying assets $S_{t,i}$ for $i \in \{1, \ldots, M\}$, along with a positive constant rate $r \geq 0$. The asset's spot prices $S_{t,i}$ evolve according to a multidimensional geometric Brownian motion:

$$dS_{t,i} = S_{t,i}\left(\mu_{i,S}dt + \sum_{k=1}^{i} \sigma_{i,k,S}dZ_{t,k}\right), \quad i \in \{1, \ldots, M\}, \tag{6.1}$$

where $\mu_{i,S}$ is the constant drift, $\sigma_{i,k,S}$, for $1 \leq k \leq i$, are constant volatility parameters, and $(Z_{t,1}, \ldots, Z_{t,M})^\top$ is a standard M-dimensional Brownian motion under the measure $\mathbb{P}$. We assume that this underlying asset or index is not continuously traded, which is typical in many futures markets. As studied by Aragon *et al.* (2020) for the VIX futures market, the futures contract and underlying are not linked by a no-arbitrage condition. Therefore, unlike Chapters 3 and 5 where the futures price is defined as a condition expectation of the future spot price, in this chapter the prices of the futures and underlying assets are linked through a multidimensional stochastic basis process.

For each underlying asset $S_{t,i}$, there are N_i futures contracts $F_{t,i,j}$ written on this asset with expiration dates $T_{i,j}$, for $j \in \{1, \ldots, N_i\}$. For counting and indexing, we define the order numbers

$$P_{i,j} = \sum_{k=1}^{i-1} N_k + j, \quad i \in \{1, \ldots, M\}, \quad j \in \{1, \ldots, N_i\},$$

and total number

$$N = \sum_{k=1}^{M} N_k = P_{M,N_M}.$$

Then, we can line up all N futures one by one, where the futures $F_{t,i,j}$ is the $P_{i,j}$-th contract.

Next, we derive the futures price dynamics via the random basis process. To that end, we define the log-value of the random basis for the futures contract $F_{t,i,j}$ by

$$Y_{t,i,j} := \log\left(\frac{F_{t,i,j}}{S_{t,i}}\right) - r(T_{i,j} - t), \tag{6.2}$$

for

$$0 \le t \le T_{i,j}, \quad i \in \{1, \dots, M\}, \quad j \in \{1, \dots, N_i\}.$$

Then, the log-basis $Y_{t,i,j}$ is assumed to evolve according to a multidimensional Brownian bridge, described by the following system of SDEs:

$$dY_{t,i,j} = \left(m_{i,j} - \frac{\kappa_{i,j} Y_{t,i,j}}{T_{i,j} - t} \right) dt + \sum_{k=1}^{P_{i,j}+M} \sigma_{P_{i,j},k,Y} dZ_{t,k}, \qquad (6.3)$$

for $i \in \{1, \dots, M\}$, $j \in \{1, \dots, N_i\}$, where drift $m_{i,j}$, coefficient $\kappa_{i,j}$ and volatility parameter $\sigma_{P_{i,j},k,Y}$ are constants for $1 \le k \le P_{i,j} + M$, and $(Z_{t,1}, \dots, Z_{t,N+M})^\top$ is a standard $N + M$ dimensional Brownian motion under the measure $\mathbb{P}$. We define the filtration $\mathbb{F} = (\mathcal{F}_t)_{t \ge 0}$ being the augmented σ-algebra generated by $\{(Z_{u,1}, \dots, Z_{u,N+M}); 0 \le u \le t\}$ and satisfies the usual conditions. By construction, each log-basis $Y_{t,i,j}$ converges to 0 at the corresponding futures maturity $T_{i,j}$.

From the basis process above, we can now derive the futures price dynamics. Precisely, using Ito's lemma, each futures price satisfies the SDE

$$dF_{t,i,j} = F_{t,i,j} \left[\left(\theta_{i,j} - \frac{\kappa_{i,j} Y_{t,i,j}}{T_{i,j} - t} \right) dt + \sum_{k=1}^{P_{i,j}+M} \sigma_{P_{i,j},k,F} dZ_{t,k} \right], \qquad (6.4)$$

where the drift parameter $\theta_{i,j}$ is given by

$$\theta_{i,j} = -r + m_{i,j} + \mu_{i,S} + \frac{1}{2} \left(2 \sum_{k=1}^{i} \sigma_{i,k,S} \sigma_{P_{i,j},k,Y} + \sum_{k=1}^{P_{i,j}+M} \sigma^2_{P_{i,j},k,Y} \right), \qquad (6.5)$$

and volatility parameter $\sigma_{P_{i,j},k,F}$ satisfies

$$\sigma_{P_{i,j},k,F} = \begin{cases} \sigma_{i,k,S} + \sigma_{P_{i,j},k,Y}, & 1 \le k \le i, \\ \sigma_{P_{i,j},k,Y}, & i < k \le P_{i,j} + M, \end{cases} \qquad (6.6)$$

for $i \in \{1, \dots, M\}$ and $j \in \{1, \dots, N_i\}$.

For a collection of futures contracts, the corresponding price dynamics follow a system of SDEs. In order to rewrite SDEs (6.1), (6.3) and (6.4) in the matrix form, we define the vector of assets $\mathbf{S}_t \in \mathbb{R}^M$, the vector of log-basis process $\mathbf{Y}_t \in \mathbb{R}^N$ and the vector of futures prices $\mathbf{F}_t \in \mathbb{R}^N$ as

$$\mathbf{S}_t := (S_{t,1}, \ldots, S_{t,M})^\top,$$
$$\mathbf{Y}_t := (Y_{t,1,1}, \ldots, Y_{t,1,N_1}, Y_{t,2,1}, \ldots, Y_{t,M,N_M})^\top,$$
$$\mathbf{F}_t := (F_{t,1,1}, \ldots, F_{t,1,N_1}, F_{t,2,1}, \ldots, F_{t,M,N_M})^\top.$$

We also define coefficients vectors and standard Brownian motions by

$$\boldsymbol{\mu} := (\mu_{1,S}, \ldots, \mu_{M,S})^\top \in \mathbb{R}^M,$$
$$\boldsymbol{\theta} := (\theta_{1,1}, \ldots, \theta_{1,N_1}, \theta_{2,1}, \ldots, \theta_{M,N_M})^\top \in \mathbb{R}^N,$$
$$\mathbf{m} := (m_{t,1,1}, \ldots, m_{t,1,N_1}, m_{t,2,1}, \ldots, m_{t,M,N_M})^\top \in \mathbb{R}^N,$$
$$\boldsymbol{K}(t) := \operatorname{diag}\left(\frac{\kappa_{1,1}}{T_{1,1}-t}, \ldots, \frac{\kappa_{1,N_1}}{T_{1,N_1}-t}, \frac{\kappa_{2,1}}{T_{2,1}-t}, \ldots, \frac{\kappa_{M,N_M}}{T_{M,N_M}-t}\right) \in \mathbb{R}^{N\times N},$$
$$\mathbf{Z}_{t,1} := (Z_{t,1}, \ldots, Z_{t,M})^\top \in \mathbb{R}^M,$$
$$\mathbf{Z}_{t,2} := (Z_{t,M+1}, \ldots, Z_{t,N+M})^\top \in \mathbb{R}^N.$$

Considering M underlying assets, we denote by $\widetilde{\boldsymbol{\Sigma}}_{\boldsymbol{S}} \in \mathbb{R}^{M\times M}$ the volatility matrix associated with the spot price process; that is,

$$\widetilde{\boldsymbol{\Sigma}}_{\boldsymbol{S}} = \begin{bmatrix} \sigma_{1,1,S} & 0 & \cdots & 0 \\ \sigma_{2,1,S} & \sigma_{2,2,S} & \cdots & 0 \\ \vdots & \vdots & \ddots & \vdots \\ \sigma_{M,1,S} & \sigma_{M,2,S} & \cdots & \sigma_{M,M,S} \end{bmatrix}.$$

Also, we define the volatility parameter matrices $\widetilde{\boldsymbol{\Sigma}}_{\boldsymbol{YS}} \in \mathbb{R}^{N\times M}$, $\widetilde{\boldsymbol{\Sigma}}_{\boldsymbol{Y}} \in \mathbb{R}^{N\times N}$ for N log-bases by

$$\widetilde{\boldsymbol{\Sigma}}_{\boldsymbol{YS}} = \begin{bmatrix} \sigma_{1,1,Y} & \cdots & \sigma_{1,M,Y} \\ \sigma_{2,1,Y} & \cdots & \sigma_{2,M,Y} \\ \vdots & \ddots & \vdots \\ \sigma_{N,1,Y} & \cdots & \sigma_{N,M,Y} \end{bmatrix},$$

$$\widetilde{\boldsymbol{\Sigma}}_{\boldsymbol{Y}} = \begin{bmatrix} \sigma_{1,M+1,Y} & 0 & \cdots & 0 \\ \sigma_{2,M+1,Y} & \sigma_{2,M+2,Y} & \cdots & 0 \\ \vdots & \vdots & \ddots & \vdots \\ \sigma_{N,M+1,Y} & \sigma_{N,M+2,Y} & \cdots & \sigma_{N,M+N,Y} \end{bmatrix},$$

and the volatility parameter matrices $\widetilde{\boldsymbol{\Sigma}}_{\boldsymbol{FS}} \in \mathbb{R}^{N\times M}$, $\widetilde{\boldsymbol{\Sigma}}_{\boldsymbol{F}} \in \mathbb{R}^{N\times N}$ for N futures by

$$\widetilde{\boldsymbol{\Sigma}}_{\boldsymbol{FS}} = \begin{bmatrix} \sigma_{1,1,F} & \cdots & \sigma_{1,M,F} \\ \sigma_{2,1,F} & \cdots & \sigma_{2,M,F} \\ \vdots & \ddots & \vdots \\ \sigma_{N,1,F} & \cdots & \sigma_{N,M,F} \end{bmatrix},$$

$$\widetilde{\boldsymbol{\Sigma}}_{\boldsymbol{F}} = \begin{bmatrix} \sigma_{1,M+1,F} & 0 & \cdots & 0 \\ \sigma_{2,M+1,F} & \sigma_{2,M+2,F} & \cdots & 0 \\ \vdots & \vdots & \ddots & \vdots \\ \sigma_{N,M+1,F} & \sigma_{N,M+2,F} & \cdots & \sigma_{N,M+N,F} \end{bmatrix}.$$

With these notations, the SDEs (6.1), (6.3), and (6.4) can be written in matrix form:

$$\begin{aligned} d\mathbf{S}_t &= \operatorname{diag}(\mathbf{S}_t)\big[\boldsymbol{\mu}dt + \widetilde{\boldsymbol{\Sigma}}_{\mathbf{S}} d\boldsymbol{Z}_{t,1}\big], \\ d\mathbf{Y}_t &= (\mathbf{m} - \boldsymbol{K}(t)\mathbf{Y}_t)dt + \widetilde{\boldsymbol{\Sigma}}_{\mathbf{YS}} d\boldsymbol{Z}_{t,1} + \widetilde{\boldsymbol{\Sigma}}_{\mathbf{Y}} d\boldsymbol{Z}_{t,2}, \\ d\mathbf{F}_t &= \operatorname{diag}(\mathbf{F}_t)\big[(\boldsymbol{\theta} - \boldsymbol{K}(t)\mathbf{Y}_t)dt + \widetilde{\boldsymbol{\Sigma}}_{\mathbf{FS}} d\boldsymbol{Z}_{t,1} + \widetilde{\boldsymbol{\Sigma}}_{\mathbf{F}} d\boldsymbol{Z}_{t,2}\big]. \end{aligned}$$

We note that by (6.6), we have $\widetilde{\boldsymbol{\Sigma}}_{\mathbf{Y}} = \widetilde{\boldsymbol{\Sigma}}_{\mathbf{F}}$ and

$$\widetilde{\boldsymbol{\Sigma}}_{\mathbf{FS}} - \widetilde{\boldsymbol{\Sigma}}_{\mathbf{YS}} = \underbrace{\begin{bmatrix} \sigma_{1,1,S} & 0 & \cdots & 0 \\ \vdots & \vdots & \ddots & \vdots \\ \sigma_{1,1,S} & 0 & \cdots & 0 \\ \sigma_{2,1,S} & \sigma_{2,2,S} & \cdots & 0 \\ \vdots & \vdots & \ddots & \vdots \\ \sigma_{2,1,S} & \sigma_{2,2,S} & \cdots & 0 \\ \vdots & \vdots & \ddots & \vdots \\ \sigma_{M,1,S} & \sigma_{M,2,S} & \cdots & \sigma_{M,M,S} \\ \vdots & \vdots & \ddots & \vdots \\ \sigma_{M,1,S} & \sigma_{M,2,S} & \cdots & \sigma_{M,M,S} \end{bmatrix}}_{M} \begin{matrix} \left.\begin{matrix} \\ \\ \\ \end{matrix}\right\} N_1 \\ \left.\begin{matrix} \\ \\ \\ \end{matrix}\right\} N_2 \\ \\ \left.\begin{matrix} \\ \\ \\ \end{matrix}\right\} N_M \end{matrix}.$$

6.2 Portfolio Optimization

Most speculative futures trading strategies involve only the futures but not the underlying assets. Indeed, many underlying assets cannot be traded directly (e.g. volatility index) or traded with sufficient liquidity (e.g. agricultural commodities). In other cases where the underlying asset is traded or accessible via an ETF (e.g. gold and silver), futures are traded as a proxy for their underlying assets because of their attractive trading characteristics such as liquidity, leverage, and the ease of taking short positions.

Motivated by these practical applications, we formulate and solve the optimal investment problem faced by a investor who invest in the market but only trades the futures contracts. This is an optimal investment problem in an incomplete market, since the underlying assets are not tradable.

To describe the futures portfolio, we let $W_t^{\boldsymbol{\pi}}$ be portfolio value over time. The amount of money invested in the futures contract $F_{t,i,j}$ is denoted by $\pi_{t,i,j}$. For simplicity, we assume a zero interest rate.

Then, the wealth process $W_t^{\boldsymbol{\pi}}$ evolves according to

$$dW_t^{\boldsymbol{\pi}} = \sum_{i=1}^{M}\sum_{j=1}^{N_i} \pi_{t,i,j}\frac{dF_{t,i,j}}{F_{t,i,j}} \tag{6.7}$$

$$= \boldsymbol{\pi}_t^\top \left[(\boldsymbol{\theta} - \boldsymbol{K}(t)\mathbf{Y}_t)dt + \widetilde{\mathbf{\Sigma}}_{\mathbf{FS}}d\boldsymbol{Z}_{t,1} + \widetilde{\mathbf{\Sigma}}_{\mathbf{F}}d\boldsymbol{Z}_{t,2}\right],$$

where we have defined the strategy as the vector

$$\boldsymbol{\pi}_t := \left(\pi_{t,1,1}, \ldots, \pi_{t,1,N_1}, \pi_{t,2,1}, \ldots, \pi_{t,M,N_M}\right)^\top.$$

In this section, we discuss the optimal trading strategy for futures portfolio with or without constraints through utility maximization. Among our findings, the associated Hamilton–Jacobi–Bellman equation (HJB) reduces to a series of ODE equations, which could be solved numerically. In addition, we provide the verification theorem for our utility maximization problem.

Like in previous chapters, the investor's risk preferences are modeled by the exponential utility

$$U(w) = -\exp(-\gamma w), \tag{6.8}$$

where $\gamma > 0$ is the risk aversion parameter.

Let's begin with general strategies without constraints, which we use superscript "no" to denote "no constraints" in our presentation.

6.2.1 *Futures portfolio without constraints*

In this section, we will discuss the case that the investor puts no portfolio constraints on the strategy $\boldsymbol{\pi}$. Then, the investor seeks an admissible strategy $\boldsymbol{\pi} \in \mathcal{A}^{no}$, that maximizes the expected utility of wealth at T, where $0 < T < T_{i,j}$ for all $i \in \{1, \dots, M\}$ and $j \in \{1, \dots, N_i\}$. It means trading stops strictly before the expiry of the futures contracts. Then, the convergence between asset's price $S_{t,i}$ and futures price $F_{t,i,j}$ is not realized in the market. This non-convergence has practical relevance since speculative futures trades are always closed out before the delivery date.

Before defining the set of admissible trading strategies, we construct an auxiliary process corresponding to any given strategy $\boldsymbol{\pi}$ by

$$A_t^{\boldsymbol{\pi}} = \int_0^t \big(- \mathbf{Y}_s^\top \boldsymbol{H}^{no}(t) + \boldsymbol{g}^{no}(t)^\top \big) \big(\widetilde{\boldsymbol{\Sigma}}_{\mathbf{Y}\mathbf{S}} d\boldsymbol{Z}_{s,1} + \widetilde{\boldsymbol{\Sigma}}_{\mathbf{Y}} d\boldsymbol{Z}_{s,2} \big)$$
$$- \gamma \boldsymbol{\pi}_s^\top \big(\widetilde{\boldsymbol{\Sigma}}_{\mathbf{F}\mathbf{S}} d\boldsymbol{Z}_{s,1} + \widetilde{\boldsymbol{\Sigma}}_{\mathbf{F}} d\boldsymbol{Z}_{s,2} \big),$$

where $\boldsymbol{H}^{no}(t), \boldsymbol{g}^{no}(t)$ are deterministic functions that depend only on the model parameters, which will appear later in Theorem 6.2 by solving the corresponding ODEs. Next, we define the set of admissible strategies.

Definition 6.1 (Admissibility). We denote $\mathcal{A}^{no}$ the set of all $\mathcal{F}_t$-adapted processes $\{\boldsymbol{\pi}_t\}_{0 \le t \le T}$, such that

(i) $\mathbb{E}^{\mathbb{P}} \left(\int_0^T |\boldsymbol{\pi}_t^\top \mathbf{Y}_t| + \|\boldsymbol{\pi}_t\|^2 dt \right) < \infty$;
(ii) $W_t^{\boldsymbol{\pi}} \in \mathcal{D}$, P-a.s., for all $t \in [0, T]$, where $\mathcal{D} = \mathbb{R}$ and $(W_t^{\boldsymbol{\pi}})_{0 \le t \le T}$ is given by (6.7);
(iii) $\mathbb{E}^{\mathbb{P}} \left(\exp \left(A_T^{\boldsymbol{\pi}} - \frac{1}{2} \langle A^{\boldsymbol{\pi}} \rangle_T \right) \right) = 1$.

Condition *(i)* is a general integrability condition to ensure the existence of the wealth process, condition *(ii)* is to assure that the wealth should be positive almost surely, and condition *(iii)* can be found in many places, e.g. Kuroda and Nagai (2002), which guarantees that the stochastic exponential of $A_t^{\boldsymbol{\pi}}$ is a martingale.

Then, the value function for the investor's utility maximization problem is defined by

$$V^{no}(t, \boldsymbol{y}, w) = \sup_{\boldsymbol{\pi} \in \mathcal{A}^{no}} \mathbb{E}^{\mathbb{P}} [U(W_T^{\boldsymbol{\pi}}) | \mathbf{Y}_t = \boldsymbol{y}, W_t^{\boldsymbol{\pi}} = w]. \tag{6.9}$$

To solve the portfolio optimization problem, we define the volatility matrices $\boldsymbol{\Sigma_Y} \in \mathbb{R}^{N\times N}$, $\boldsymbol{\Sigma_{FY}} \in \mathbb{R}^{N\times N}$ and $\boldsymbol{\Sigma_F} \in \mathbb{R}^{N\times N}$ matrix, as

$$\boldsymbol{\Sigma_Y} = \widetilde{\boldsymbol{\Sigma}}_{\mathbf{YS}}\widetilde{\boldsymbol{\Sigma}}_{\mathbf{YS}}^\top + \widetilde{\boldsymbol{\Sigma}}_{\mathbf{Y}}\widetilde{\boldsymbol{\Sigma}}_{\mathbf{Y}}^\top, \tag{6.10}$$

$$\boldsymbol{\Sigma_{FY}} = \widetilde{\boldsymbol{\Sigma}}_{\mathbf{FS}}\widetilde{\boldsymbol{\Sigma}}_{\mathbf{YS}}^\top + \widetilde{\boldsymbol{\Sigma}}_{\mathbf{F}}\widetilde{\boldsymbol{\Sigma}}_{\mathbf{Y}}^\top, \tag{6.11}$$

$$\boldsymbol{\Sigma_F} = \widetilde{\boldsymbol{\Sigma}}_{\mathbf{FS}}\widetilde{\boldsymbol{\Sigma}}_{\mathbf{FS}}^\top + \widetilde{\boldsymbol{\Sigma}}_{\mathbf{F}}\widetilde{\boldsymbol{\Sigma}}_{\mathbf{F}}^\top. \tag{6.12}$$

Also to facilitate the presentation, we define the linear operator

$$\mathcal{L}\,\cdot = (\mathbf{m} - \boldsymbol{K}(t)\boldsymbol{y})^\top \nabla_{\boldsymbol{y}} \cdot + \frac{1}{2}\mathrm{Tr}(\boldsymbol{\Sigma_Y}\nabla_{\boldsymbol{y}}^2 \cdot), \tag{6.13}$$

where

$$\nabla_{\boldsymbol{y}}\cdot = (\partial_{y_{1,1}}\cdot, \ldots, \partial_{y_{1,N_1}}\cdot, \partial_{y_{2,1}}\cdot, \ldots, \partial_{y_{M,N_M}}\cdot)^\top$$

is the nabla operator, and the Hessian operator $\nabla_{\boldsymbol{y}}^2\cdot$ satisfies

$$\nabla_{\boldsymbol{y}}^2\cdot = \begin{bmatrix} \partial_{y_{1,1}}^2\cdot & \partial_{y_{1,1}}\partial_{y_{1,2}}\cdot & \cdots & \partial_{y_{1,1}}\partial_{y_{N,N_M}}\cdot \\ \partial_{y_{1,2}}\partial_{y_{1,1}}\cdot & \partial_{y_{1,2}}^2\cdot & \cdots & \partial_{y_{1,2}}\partial_{y_{N,N_M}}\cdot \\ \vdots & \vdots & \ddots & \vdots \\ \partial_{y_{N,N_M}}\partial_{y_{1,1}}\cdot & \partial_{y_{N,N_M}}\partial_{y_{1,2}}\cdot & \cdots & \partial_{y_{N,N_M}}^2\cdot \end{bmatrix}.$$

and column-valued-function $\boldsymbol{a}^{no}(t, \boldsymbol{y}, w)$

$$\boldsymbol{a}^{no}(t, \boldsymbol{y}, w) = (\boldsymbol{\theta} - \boldsymbol{K}(t)\boldsymbol{y})\partial_w u^{no} + \boldsymbol{\Sigma_{FY}}\nabla_{\boldsymbol{y}}\partial_w u^{no}.$$

With the above definitions, we now state the HJB equation for the candidate value function $u^{no}(t, \boldsymbol{y}, w)$,

$$\partial_t u^{no} + \mathcal{L}u^{no} + \max_{\boldsymbol{\pi}\in\mathcal{A}}\left\{\boldsymbol{\pi}^\top \boldsymbol{a}^{no}(t, \boldsymbol{y}, w) + \frac{\partial_{ww}u^{no}}{2}\boldsymbol{\pi}^\top\boldsymbol{\Sigma_F}\boldsymbol{\pi}\right\} = 0, \tag{6.14}$$

for $(t, \boldsymbol{y}, w) \in [0, T) \times \mathbb{R}^N \times \mathcal{D}$, where operator $\mathcal{L}$ is defined in (6.13). The terminal condition is

$$u^{no}(T, \boldsymbol{y}, w) = U(w) = -e^{-\gamma w},$$

for $(\boldsymbol{y}, w) \in \mathbb{R}^N \times \mathcal{D}$.

Next, we present the solution to the HJB equation (6.14). To that end, we first define

$$\boldsymbol{\Sigma}^{no} = \boldsymbol{\Sigma}_{\boldsymbol{Y}} - \boldsymbol{\Sigma}_{\boldsymbol{FY}}^{\top}\boldsymbol{\Sigma}_{\boldsymbol{F}}^{-1}\boldsymbol{\Sigma}_{\boldsymbol{FY}}, \tag{6.15}$$

which is a $N \times N$ matrix.

Theorem 6.2.

(1) *The matrix Riccati differential equation below has a unique solution that is positive definite for all* $t \in [0, T]$,

$$\begin{aligned}
\frac{d}{dt}\boldsymbol{H}^{no}(t) &= \left(\boldsymbol{K}(t) - \boldsymbol{K}(t)\boldsymbol{\Sigma}_{\boldsymbol{F}}^{-1}\boldsymbol{\Sigma}_{\boldsymbol{FY}}\right)\boldsymbol{H}^{no}(t)\\
&\quad + \boldsymbol{H}^{no}(t)\left(\boldsymbol{K}(t) - \boldsymbol{K}(t)\boldsymbol{\Sigma}_{\boldsymbol{F}}^{-1}\boldsymbol{\Sigma}_{\boldsymbol{FY}}\right)^{\top} \qquad (6.16)\\
&\quad + \boldsymbol{H}^{no}(t)\boldsymbol{\Sigma}^{no}\boldsymbol{H}^{no}(t) - \boldsymbol{K}(t)^{\top}\boldsymbol{\Sigma}_{\boldsymbol{F}}^{-1}\boldsymbol{K}(t),\\
\boldsymbol{H}^{no}(T) &= \boldsymbol{0}_{N\times N},
\end{aligned}$$

where $\boldsymbol{0}_{N\times N}$ *denotes the zero matrix of dimension* $N \times N$.

(2) *The solution of the HJB equation (6.14) is given by*

$$u^{no}(t, \boldsymbol{y}, w) = U(w)\exp\left(-\frac{1}{2}\boldsymbol{y}^{\top}\boldsymbol{H}^{no}(t)\boldsymbol{y} + \boldsymbol{y}^{\top}\boldsymbol{g}^{no}(t) + f^{no}(t)\right), \tag{6.17}$$

for $(t, \boldsymbol{y}, w) \in [0, T] \times \mathbb{R}^N \times \mathbb{R}^+$, *where* $\boldsymbol{H}^{no}(t) \in \mathbb{R}^{N\times N}$ *satisfies the matrix Riccati differential equation (6.16),* $\boldsymbol{g}^{no}(t) \in \mathbb{R}^N$ *satisfies the following ODE system:*

$$\begin{aligned}
\frac{d}{dt}\boldsymbol{g}^{no}(t) &= \boldsymbol{K}(t)\boldsymbol{g}^{no}(t) + \boldsymbol{H}^{no}(t)\boldsymbol{m}\\
&\quad + \left(\boldsymbol{H}^{no}(t)\boldsymbol{\Sigma}^{no} - \boldsymbol{K}(t)\boldsymbol{\Sigma}_{\boldsymbol{F}}^{-1}\boldsymbol{\Sigma}_{\boldsymbol{FY}}\right)\boldsymbol{g}^{no}(t)\\
&\quad - \left(\boldsymbol{K}(t) + \boldsymbol{H}^{no}(t)\boldsymbol{\Sigma}_{\boldsymbol{FY}}^{\top}\right)\boldsymbol{\Sigma}_{\boldsymbol{F}}^{-1}\boldsymbol{\theta}, \qquad (6.18)\\
\boldsymbol{g}^{no}(T) &= \boldsymbol{0}_{N\times 1},
\end{aligned}$$

and $f^{no}(t) \in \mathbb{R}$ *follows the ODE:*

$$\begin{aligned}
\frac{d}{dt}f^{no}(t) &= -\boldsymbol{m}^{\top}\boldsymbol{g}^{no}(t) + \frac{1}{2}tr\left(\boldsymbol{\Sigma}_{\boldsymbol{Y}}\boldsymbol{H}^{no}(t)\right) - \frac{1}{2}\boldsymbol{g}^{no}(t)^{\top}\boldsymbol{\Sigma}_{\boldsymbol{Y}}\boldsymbol{g}^{no}(t)\\
&\quad + \frac{1}{2}\left(\boldsymbol{\theta} + \boldsymbol{\Sigma}_{\boldsymbol{FY}}\boldsymbol{g}^{no}(t)\right)^{\top}\boldsymbol{\Sigma}_{\boldsymbol{F}}^{-1}\left(\boldsymbol{\theta} + \boldsymbol{\Sigma}_{\boldsymbol{FY}}\boldsymbol{g}^{no}(t)\right), \qquad (6.19)\\
f^{no}(T) &= 0.
\end{aligned}$$

(3) *The optimal strategy is given by*

$$\boldsymbol{\pi}^*(t,\boldsymbol{y}) = \frac{1}{\gamma}\boldsymbol{\Sigma}_{\boldsymbol{F}}^{-1}\big(\boldsymbol{\theta} - \boldsymbol{K}(t)\boldsymbol{y} + \boldsymbol{\Sigma}_{\boldsymbol{FY}}\big(\boldsymbol{g}^{no}(t) - \boldsymbol{H}^{no}(t)\boldsymbol{y}\big)\big). \tag{6.20}$$

Proof. A detailed proof is provided in Section 6.6. □

Futures portfolios often come with portfolio constraints, among which dollar neutrality or market neutrality are the most popular. Within the stochastic basis framework, we now give a rigorous definition of portfolio constraints, and introduce a general class of constraints that includes both dollar and market neutrality.

6.2.2 *Constrained futures portfolio*

We now introduce portfolio constraints to the futures trading problem. The portfolio constrains considered in this model are linear. They can be represented in matrix form, given by

$$\boldsymbol{\Gamma}^\top \boldsymbol{\pi} = \boldsymbol{c}, \tag{6.21}$$

where $\boldsymbol{\Gamma}$ is a $N \times d$ matrix, and $\boldsymbol{c} \in \mathbb{R}^d$ is a column vector.

For practical reasons, we assume these d constraints are linearly independent, i.e. $\text{rank}(\boldsymbol{\Gamma}) = d$. Otherwise, some constraints are either redundant or infeasible.

The admissible set $\mathcal{A}$ for constrained case is almost the same that in Definition 6.1, except that $\boldsymbol{H}^{no}$ and $\boldsymbol{g}^{no}$ in the auxiliary process $A^{\boldsymbol{\pi}}$ should be replaced by $\boldsymbol{H}$ and $\boldsymbol{g}$ (given in (6.27) and (6.29)) respectively. With this, the investor is assumed to seek an admissible strategy within the set defined by

$$\{\boldsymbol{\pi} \in \mathcal{A} \,|\, \boldsymbol{\Gamma}^\top \boldsymbol{\pi}_t = \boldsymbol{c}, \quad \text{for } t \in [0,T]\}.$$

The investor select an admissible strategy to maximize the expected utility of wealth at T. Therefore, the value function is defined as

$$V(t,\boldsymbol{y},w) = \sup_{\boldsymbol{\pi}\in\mathcal{A},\boldsymbol{\Gamma}^\top\boldsymbol{\pi}=\boldsymbol{c}} \mathbb{E}^{\mathbb{P}}[U(W_T^{\boldsymbol{\pi}})|\boldsymbol{Y}_t = \boldsymbol{y}, W_t^{\boldsymbol{\pi}} = w]. \tag{6.22}$$

In order to derive the corresponding HJB equation for the value function, we first define the column-vector-valued function

$$\boldsymbol{a}(t,\boldsymbol{y},w) = (\partial_w u)(\boldsymbol{\theta} - \boldsymbol{K}(t)\boldsymbol{y}) + \boldsymbol{\Sigma}_{\boldsymbol{FY}}(\partial_w \nabla_{\boldsymbol{y}} u).$$

Then, we write down the HJB equation for the candidate value function $u(t, \boldsymbol{y}, w)$:

$$\partial_t u + \mathcal{L} u + \max_{\boldsymbol{\pi} \in \mathcal{A}, \boldsymbol{\Gamma}^\top \boldsymbol{\pi} = \boldsymbol{c}} \left\{ \boldsymbol{\pi}^\top \boldsymbol{a}(t, \boldsymbol{y}, w) + \frac{\partial_{ww} u}{2} \boldsymbol{\pi}^\top \boldsymbol{\Sigma}_{\boldsymbol{F}} \boldsymbol{\pi} \right\} = 0, \qquad (6.23)$$

for $(t, \boldsymbol{y}, w) \in [0, T) \times \mathbb{R}^N \times \mathcal{D}$, where the set $\mathcal{D}$ contains all possible values for the wealth w. The terminal condition is

$$u(T, \boldsymbol{y}, w) = U(w),$$

for $(\boldsymbol{y}, w) \in \mathbb{R}^N \times \mathcal{D}$.

Now we present the solution to the HJB equation (6.23). To that end, we first define $\boldsymbol{D}_{\boldsymbol{\Gamma}} \in \mathbb{R}^{d \times d}$, $\boldsymbol{\Sigma}_{\boldsymbol{\Gamma}} \in \mathbb{R}^{N \times N}$ and $\boldsymbol{\Sigma} \in \mathbb{R}^{N \times N}$ as

$$\boldsymbol{D}_{\boldsymbol{\Gamma}} = \boldsymbol{\Gamma}^\top \boldsymbol{\Sigma}_{\boldsymbol{F}}^{-1} \boldsymbol{\Gamma}, \qquad (6.24)$$

$$\boldsymbol{\Sigma}_{\boldsymbol{\Gamma}} = \boldsymbol{\Sigma}_{\boldsymbol{F}}^{-1} \boldsymbol{\Gamma} \boldsymbol{D}_{\boldsymbol{\Gamma}}^{-1} \boldsymbol{\Gamma}^\top \boldsymbol{\Sigma}_{\boldsymbol{F}}^{-1}, \qquad (6.25)$$

$$\boldsymbol{\Sigma} = \boldsymbol{\Sigma}_{\boldsymbol{Y}} - \boldsymbol{\Sigma}_{\boldsymbol{F}\boldsymbol{Y}}^\top (\boldsymbol{\Sigma}_{\boldsymbol{F}}^{-1} - \boldsymbol{\Sigma}_{\boldsymbol{\Gamma}}) \boldsymbol{\Sigma}_{\boldsymbol{F}\boldsymbol{Y}}. \qquad (6.26)$$

Note that $\boldsymbol{D}_{\boldsymbol{\Gamma}}$ is invertible due to the assumption that $\operatorname{rank}(\boldsymbol{\Gamma}) = d$.

Theorem 6.3. *Assume the investor's utility is given by (6.8). Then, the following statements hold.*

(1) *The matrix Riccati differential equation below has a unique solution that is positive definite for all $t \in [0, T]$,*

$$\begin{aligned} \boldsymbol{H}'(t) &= \left(\boldsymbol{K}(t) - \boldsymbol{K}(t)(\boldsymbol{\Sigma}_{\boldsymbol{F}}^{-1} - \boldsymbol{\Sigma}_{\boldsymbol{\Gamma}})\boldsymbol{\Sigma}_{\boldsymbol{F}\boldsymbol{Y}}\right) \boldsymbol{H}(t) \\ &\quad + \boldsymbol{H}(t) \left(\boldsymbol{K}(t) - \boldsymbol{K}(t)(\boldsymbol{\Sigma}_{\boldsymbol{F}}^{-1} - \boldsymbol{\Sigma}_{\boldsymbol{\Gamma}})\boldsymbol{\Sigma}_{\boldsymbol{F}\boldsymbol{Y}}\right)^\top \qquad (6.27) \\ &\quad + \boldsymbol{H}(t) \boldsymbol{\Sigma} \boldsymbol{H}(t) - \boldsymbol{K}(t)(\boldsymbol{\Sigma}_{\boldsymbol{F}}^{-1} - \boldsymbol{\Sigma}_{\boldsymbol{\Gamma}})\boldsymbol{K}(t), \\ \boldsymbol{H}(T) &= \boldsymbol{0}_{N \times N}, \end{aligned}$$

where $\boldsymbol{0}_{N \times N}$ denotes the zero matrix of dimension $N \times N$.

(2) *The solution of the HJB equation (6.23) is given by*

$$u(t, \boldsymbol{y}, w) = U(w) \exp\left(-\frac{1}{2} \boldsymbol{y}^\top \boldsymbol{H}(t) \boldsymbol{y} + \boldsymbol{y}^\top \boldsymbol{g}(t) + f(t) \right), \qquad (6.28)$$

for $(t, \boldsymbol{y}, w) \in [0, T] \times \mathbb{R}^N \times \mathbb{R}^+$, where $\boldsymbol{H}(t) \in \mathbb{R}^{N \times N}$ satisfies the matrix Riccati differential equation (6.27), $\boldsymbol{g}(t) \in \mathbb{R}^N$ satisfies the following

ODE system:

$$\begin{aligned} \boldsymbol{g}'(t) &= \left(\boldsymbol{K}(t) + \boldsymbol{H}(t)\boldsymbol{\Sigma} - \boldsymbol{K}(t)(\boldsymbol{\Sigma}_{\boldsymbol{F}}^{-1} - \boldsymbol{\Sigma}_{\Gamma})\boldsymbol{\Sigma}_{\boldsymbol{F}\boldsymbol{Y}}\right)\boldsymbol{g}(t) + \boldsymbol{H}(t)\boldsymbol{m} \\ &\quad - \left(\boldsymbol{K}(t) + \boldsymbol{H}(t)\boldsymbol{\Sigma}_{\boldsymbol{F}\boldsymbol{Y}}^{\top}\right)\left((\boldsymbol{\Sigma}_{\boldsymbol{F}}^{-1} - \boldsymbol{\Sigma}_{\Gamma})\boldsymbol{\theta} + \gamma\boldsymbol{\Sigma}_{\boldsymbol{F}}^{-1}\boldsymbol{\Gamma}\boldsymbol{D}_{\Gamma}^{-1}\boldsymbol{c}\right), \quad (6.29) \\ \boldsymbol{g}(T) &= \boldsymbol{0}_{N\times 1}, \end{aligned}$$

and $f(t) \in \mathbb{R}$ follows the ODE,

$$\begin{aligned} f'(t) &= -\boldsymbol{m}^{\top}\boldsymbol{g}(t) + \frac{1}{2}Tr(\boldsymbol{\Sigma}_{\boldsymbol{Y}}\boldsymbol{H}(t)) - \frac{1}{2}\boldsymbol{g}(t)^{\top}\boldsymbol{\Sigma}_{\boldsymbol{Y}}\boldsymbol{g}(t) \\ &\quad + \frac{1}{2}\left(\boldsymbol{\theta} + \boldsymbol{\Sigma}_{\boldsymbol{F}\boldsymbol{Y}}\boldsymbol{g}(t)\right)^{\top}\left(\boldsymbol{\Sigma}_{\boldsymbol{F}}^{-1} - \boldsymbol{\Sigma}_{\Gamma}\right)\left(\boldsymbol{\theta} + \boldsymbol{\Sigma}_{\boldsymbol{F}\boldsymbol{Y}}\boldsymbol{g}(t)\right) \quad (6.30) \\ &\quad + \gamma\left(\boldsymbol{\theta} + \boldsymbol{\Sigma}_{\boldsymbol{F}\boldsymbol{Y}}\boldsymbol{g}(t)\right)^{\top}\boldsymbol{\Sigma}_{\boldsymbol{F}}^{-1}\boldsymbol{\Gamma}\boldsymbol{D}_{\Gamma}^{-1}\boldsymbol{c} - \frac{\gamma^2}{2}\boldsymbol{c}^{\top}\boldsymbol{D}_{\Gamma}^{-1}\boldsymbol{c}, \\ f(T) &= 0. \end{aligned}$$

(3) *The optimal strategy is given by*

$$\boldsymbol{\pi}^*(t, \boldsymbol{y}) = \boldsymbol{\Sigma}_{\boldsymbol{F}}^{-1}\boldsymbol{\Gamma}\boldsymbol{D}_{\Gamma}^{-1}\boldsymbol{c} + \frac{1}{\gamma}(\boldsymbol{\Sigma}_{\boldsymbol{F}}^{-1} - \boldsymbol{\Sigma}_{\Gamma})\left(\boldsymbol{\theta} - \boldsymbol{K}(t)\boldsymbol{y} + \boldsymbol{\Sigma}_{\boldsymbol{F}\boldsymbol{Y}}\left(\boldsymbol{g}(t) - \boldsymbol{H}(t)\boldsymbol{y}\right)\right). \quad (6.31)$$

Proof. See Section 6.6. □

Remark 6.4. We have considered equality constraints. If instead inequality constraints of the form

$$\boldsymbol{\Gamma}^{\top}\boldsymbol{\pi} \leq \boldsymbol{c}$$

are imposed in (6.23), then one must maximize the Lagrangian function (6.35) with constraints $\boldsymbol{\lambda}(t) \geq \boldsymbol{0}$. There is no analytic solution to this constrained quadratic optimization problem, so numerical methods need to be developed for this extension, which is not discussed in this book.

Let's look at the optimal strategies associated with some well-known portfolio constraints.

Example 6.5 (Dollar Neutral Strategy). A dollar neutral strategy invests the same amount of money in long and short positions. The resulting portfolio is certainly not risk-free and end up positively or negatively correlated with the underlying assets. Mathematically, a portfolio is said to be dollar neutral if the corresponding strategy satisfies the single constraint:

$$\mathbf{1}_{N\times 1}^{\top}\boldsymbol{\pi} = 0,$$

where $\mathbf{1}_{N\times 1}$ is a vector of N ones. This amounts to set $\boldsymbol{c} = 0$ (scalar) and $\boldsymbol{\Gamma} = \mathbf{1}_{N\times 1}$ in (6.21). When $N = 1$, meaning that only one futures is traded, the strategy $\boldsymbol{\pi}$ has to be zero to satisfy the dollar neutral constraint.

Then, according to Theorem 6.3, the optimal strategy for dollar neutral portfolio is given by

$$\boldsymbol{\pi}_t^* = \frac{1}{\gamma}(\boldsymbol{\Sigma}_{\boldsymbol{F}}^{-1} - \boldsymbol{\Sigma}_{\Gamma})(\boldsymbol{\theta} - \boldsymbol{K}(t)\boldsymbol{y} + \boldsymbol{\Sigma}_{\boldsymbol{FY}}(\boldsymbol{g}(t) - \boldsymbol{H}(t)\boldsymbol{y})).$$

Compared to the optimal strategy in (6.31), which is a sum of two terms, the dollar neutral strategy is simpler.

Example 6.6 (Market Neutral Strategy). A market neutral strategy is designed to eliminate the correlation between the portfolio and market (see Patton (2008); Valle and N. Meade (2014)). For futures portfolios, a trading strategy is said to be market neutral if the resulting portfolio wealth process $W_t^{\boldsymbol{\pi}}$ is uncorrelated with the underlying assets $S_{i,t}$, satisfying

$$dW_t^{\boldsymbol{\pi}}\, dS_{i,t} = 0,$$

for $i \in \{1, \ldots, M\}$.

Hence, the market neutral constraints are described by the system of equations:

$$(\widetilde{\boldsymbol{\Sigma}}_{\boldsymbol{FS}}\widetilde{\boldsymbol{\Sigma}}_{\boldsymbol{S}}^{\top})^{\top}\boldsymbol{\pi} = \mathbf{0}_{M\times 1}.$$

This amounts to setting in (6.21)

$$\boldsymbol{\Gamma} = \widetilde{\Sigma}_{\boldsymbol{FS}}\widetilde{\Sigma}_{\boldsymbol{S}}^{\top}, \quad \text{and} \quad \boldsymbol{c} = \mathbf{0}.$$

In particular, if $\boldsymbol{\Gamma}$ is an invertible matrix, then it follows that $\boldsymbol{\pi}_t^* = \mathbf{0}_{N\times 1}$, which means that the only feasible strategy is to not trade anything at all. This issue can be remedied by increasing the number of futures traded (see Figure 6.3 and its discussion).

Then according to Theorem 6.3, the optimal strategy for the market neutral portfolio is given by

$$\boldsymbol{\pi}_t^* = \frac{1}{\gamma}(\boldsymbol{\Sigma}_{\boldsymbol{F}}^{-1} - \boldsymbol{\Sigma}_{\Gamma})(\boldsymbol{\theta} - \boldsymbol{K}(t)\boldsymbol{y} + \boldsymbol{\Sigma}_{\boldsymbol{FY}}(\boldsymbol{g}(t) - \boldsymbol{H}(t)\boldsymbol{y})).$$

When the number of futures is less than or equal to M, the number of assets, the rank of matrix $\widetilde{\boldsymbol{\Sigma}}_{\boldsymbol{FS}}\widetilde{\boldsymbol{\Sigma}}_{\boldsymbol{S}}^{\top}$ will be less than or equal to M, which yields $\boldsymbol{\pi} = \mathbf{0}$.

Market neutral constraint is taken into consideration to determine optimal pair trading in Liu and Timmermann (2013) and Angoshtari (2016), while Zhao and Palomar (2018) propose a mean-reverting portfolio design with an investment budget constraint. Most recently, Li and Papanicolaou (2019) analyze the optimal portfolio for multiple co-integrated assets with a general linear constraint. Compared to these studies, our model is a multidimensional stochastic basis framework for futures trading with portfolio constraints. Our analysis show how optimal strategies and value function depend on the stochastic basis and different portfolio constraints.

The dollar neutral and market neutral constraints are often used in practice. Since the functions $\{\boldsymbol{\Sigma}_{\boldsymbol{\Gamma}}, \boldsymbol{H}(t), \boldsymbol{g}(t)\}$ all depend on the choice of portfolio constraints, $\boldsymbol{\Gamma}$ and $\boldsymbol{c}$, the optimal strategies presented in two examples above are different.

In order to unify the two special constraints above, we introduce a class of constraints that enforces neutrality with respect to any given constraint matrix Γ.

Definition 6.7 ($\boldsymbol{\Gamma}$-Neutral). A strategy $\boldsymbol{\pi}$ is said to be $\boldsymbol{\Gamma}$-*neutral* if it satisfies the constraints:

$$\boldsymbol{\Gamma}^{\top}\boldsymbol{\pi} = \mathbf{0}.$$

Remark 6.8. Recall that $\boldsymbol{\pi}$ stands for the amount money invested, but the $\boldsymbol{\Gamma}$-neutral condition also holds for portfolio weights.

With the definition of $\boldsymbol{\Gamma}$-neutral strategy, we can now view the dollar neutral strategy as setting $\boldsymbol{\Gamma} = \mathbf{1}$ and the market neutral strategy as setting $\widetilde{\boldsymbol{\Sigma}}_{\boldsymbol{SF}}\widetilde{\boldsymbol{\Sigma}}_{\boldsymbol{S}}^{\top}$. Moreover, we can decompose the optimal strategies for general constraints $\boldsymbol{\Gamma}^{\top}\boldsymbol{\pi} = \boldsymbol{c}$ into two components, one of which is dominated by the $\boldsymbol{\Gamma}$-neutral case, and the remaining component reveals the hedging demand when $\boldsymbol{c} \neq 0$. To that end, we first decompose the coefficient functions $\boldsymbol{g}(t)$ and $f(t)$ as follows:

Lemma 6.9. *There exist $\boldsymbol{\Psi}(t)$, $\boldsymbol{\beta}(t)$, $\boldsymbol{\Lambda}(t)$, which are $N \times d, 1 \times d$ and $d \times d$ matrices, respectively, such that*

$$\begin{aligned} \boldsymbol{g}(t) &= \boldsymbol{g}_0(t) + \boldsymbol{\Psi}(t)\boldsymbol{c}, \\ f(t) &= f_0(t) + \boldsymbol{\beta}(t)\boldsymbol{c} + \boldsymbol{c}^{\top}\boldsymbol{\Lambda}(t)\boldsymbol{c}, \end{aligned} \tag{6.32}$$

where $g_0(t)$ *and* $f_0(t)$ *correspond to the solutions of ODEs* (6.29) *and* (6.30) *under the* Γ*-neutral constraint* $\mathbf{\Gamma}^\top \boldsymbol{\pi} = \mathbf{0}$. *In addition,* $\mathbf{\Lambda}(t)$ *is positive semidefinite.*

Proof. The detailed proof is provided in Section 6.6. □

While $\boldsymbol{g}(t)$ and $f(t)$ are found from solving an ODE system and may not be fully explicit, the decomposition in Lemma 6.9 reveals some interesting structure in terms of the effects of $\boldsymbol{c}$. Since the functions $\boldsymbol{\Psi}(t)$, $\boldsymbol{\beta}(t)$, and $\mathbf{\Lambda}(t)$ are all independent of $\boldsymbol{c}$. Additionally, given that the ODEs for $\boldsymbol{H}(t)$ do not depend on $\boldsymbol{c}$, $\boldsymbol{H}(t)$ is the same as that in $\mathbf{\Gamma}$-neutral case. Hence, applying Lemma 6.9 and rearranging (6.31), we obtain the following decomposition for optimal strategy.

Proposition 6.10. *The optimal strategy in* (6.31) *can be decomposed into two components*:

$$\begin{aligned}\boldsymbol{\pi}^*(t,\boldsymbol{y}) &= \boldsymbol{\Sigma}_{\boldsymbol{F}}^{-1}\mathbf{\Gamma}\boldsymbol{D}_{\Gamma}^{-1}\boldsymbol{c} + \frac{1}{\gamma}(\boldsymbol{\Sigma}_{\boldsymbol{F}}^{-1} - \boldsymbol{\Sigma}_{\Gamma})(\boldsymbol{\theta} - \boldsymbol{K}(t)\boldsymbol{y} + \boldsymbol{\Sigma}_{\boldsymbol{FY}}(\boldsymbol{g}(t) - \boldsymbol{H}(t)\boldsymbol{y})),\\ &= \underbrace{\frac{1}{\gamma}(\boldsymbol{\Sigma}_{\boldsymbol{F}}^{-1} - \boldsymbol{\Sigma}_{\Gamma})(\boldsymbol{\theta} - \boldsymbol{K}(t)\boldsymbol{y} + \boldsymbol{\Sigma}_{\boldsymbol{FY}}(\boldsymbol{g}_0(t) - \boldsymbol{H}(t)\boldsymbol{y}))}_{\mathbf{\Gamma}\text{-neutral holding position}}\\ &\quad + \underbrace{\left(\boldsymbol{\Sigma}_{\boldsymbol{F}}^{-1}\mathbf{\Gamma}\boldsymbol{D}_{\Gamma}^{-1} + \frac{1}{\gamma}(\boldsymbol{\Sigma}_{\boldsymbol{F}}^{-1} - \boldsymbol{\Sigma}_{\Gamma})\boldsymbol{\Sigma}_{\boldsymbol{FY}}\boldsymbol{\Phi}(t)\right)\boldsymbol{c}}_{\text{hedging demand for } \boldsymbol{c}}.\end{aligned}$$

The above proposition gives us a useful decomposition of the optimal strategy. It also illustrates the dependence of the strategy on the constraint.

The first component of the optimal strategy reflects the portion due to the $\mathbf{\Gamma}$-neutral constraint. In other words, we can view the optimal strategy as the optimal $\mathbf{\Gamma}$-neutral portfolio plus something more. The additional position comes from the second component, which has linear dependence on $\boldsymbol{c}$. We call the second term the *hedging demand* as it represents extra positions required due to a non-zero constraint vector $\boldsymbol{c}$.

In general, any admissible strategy in the constrained portfolio case is also admissible in unconstrained case, so one expects the value function for the unconstrained portfolio to dominate that for the constrained one. Furthermore, by directly comparing (6.17) and (6.28), we can in fact find a direct link between the value functions by analyzing the associated HJB equations and solutions $u(t,\boldsymbol{y},w)$ and $u^{no}(t,\boldsymbol{y},w)$.

Proposition 6.11. *Define the auxiliary functions:*

$$\widetilde{\boldsymbol{H}}(t) = \boldsymbol{H}^{no}(t) - \boldsymbol{H}(t),$$
$$\widetilde{\boldsymbol{g}}(t) = \boldsymbol{g}^{no}(t) - \boldsymbol{g}(t),$$
$$\widetilde{f}(t) = f^{no}(t) - f(t).$$

Then, we have $u(t, \boldsymbol{y}, w)$ *and* $u^{no}(t, \boldsymbol{y}, w)$*:*

$$u(t, \boldsymbol{y}, w) = u^{no}(t, \boldsymbol{y}, w) \exp\left(\frac{1}{2}\left(\boldsymbol{y}^\top, 1\right)\begin{pmatrix} \widetilde{\boldsymbol{H}}(t), & -\widetilde{\boldsymbol{g}}(t) \\ -\widetilde{\boldsymbol{g}}(t)^\top, & -2\widetilde{f}(t) \end{pmatrix}\begin{pmatrix} \boldsymbol{y} \\ 1 \end{pmatrix}\right). \tag{6.33}$$

In addition, the matrix $\begin{pmatrix} \widetilde{\boldsymbol{H}}(t), & -\widetilde{\boldsymbol{g}}(t) \\ -\widetilde{\boldsymbol{g}}(t)^\top, & -2\widetilde{f}(t) \end{pmatrix}$ *is positive semidefinite.*

Proof. See Section 6.6. □

Since the matrix

$$\begin{pmatrix} \widetilde{\boldsymbol{H}}(t), & -\widetilde{\boldsymbol{g}}(t) \\ -\widetilde{\boldsymbol{g}}(t)^\top, & -2\widetilde{f}(t) \end{pmatrix}$$

is positive semidefinite by Proposition 6.11, the exponential term in (6.33) is greater than 1. Couple this with the fact that the value functions $u^{no}(t, \boldsymbol{y}, w)$ and $u(t, \boldsymbol{y}, w)$ are negative, we have the following result.

Corollary 6.12. *We have inequality*

$$u(t, \boldsymbol{y}, w) \leq u^{no}(t, \boldsymbol{y}, w).$$

Next, we verify that the value function (6.22) coincides with the solution of HJB equation (6.23) from Theorem 6.3. We also identify the optimal trading strategy.

Theorem 6.13 (Verification Theorem).

(1) *The value function in* (6.9) *is equal to the function* u^{no} *given in Theorem* 6.2. *Furthermore, the optimal trading strategy is given by* (6.20).
(2) *The value function in* (6.22) *is equal to the function* u *given in Theorem* 6.3. *Furthermore, the optimal trading strategy is given by* (6.31).

Proof. Since the unconstrained scenario is just a special constrained case where $\boldsymbol{\Gamma} = \boldsymbol{0}$ and $\boldsymbol{c} = \boldsymbol{0}$, it suffices to prove the second statement. See Section 6.6. □

6.3 Certainty Equivalent

In order to quantify the value of investing in a dynamic futures portfolio, we utilize the notion of certainty equivalent (CE). Certainty equivalent is the guaranteed cash amount that would yield the same utility as that from optimally and dynamically trading futures. This amounts to applying the inverse of the exponential utility function to the value function associated with the futures trading problem.

Since there are several cases with and without constraints, let us denote $CE(t, \boldsymbol{y}, w)$ to be the certainty equivalent value for the general constrained portfolio, and $CE^{no}(t, \boldsymbol{y}, w)$ for the unconstrained case. They are defined as follows:

$$CE(t, \boldsymbol{y}, w) = w + \frac{1}{\gamma}\left(\frac{1}{2}\boldsymbol{y}^\top \boldsymbol{H}(t)\boldsymbol{y} - \boldsymbol{y}^\top \boldsymbol{g}(t) - f(t)\right),$$

and

$$CE^{no}(t, \boldsymbol{y}, w) = w + \frac{1}{\gamma}\left(\frac{1}{2}\boldsymbol{y}^\top \boldsymbol{H}^{no}(t)\boldsymbol{y} - \boldsymbol{y}^\top \boldsymbol{g}^{no}(t) - f^{no}(t)\right).$$

In particular, for the $\boldsymbol{\Gamma}-$neutral case, the certainty equivalent is defined by

$$CE_0(t, \boldsymbol{y}, w) = w + \frac{1}{\gamma}\left(\frac{1}{2}\boldsymbol{y}^\top \boldsymbol{H}_0(t)\boldsymbol{y} - \boldsymbol{y}^\top \boldsymbol{g}_0(t) - f_0(t)\right),$$

where $f_0, \boldsymbol{g}_0$ are given in Lemma 6.9 and $H_0(t) = H(t)$.

In Theorem 6.11, we compare the unconstrained and constrained portfolio optimization problems through the corresponding value functions. We can now do the same using certainty equivalents.

Proposition 6.14. *The constrained and unconstrained cases are connected through the equality*

$$CE(t, \boldsymbol{y}, w) = CE^{no}(t, \boldsymbol{y}, w) - \frac{1}{2\gamma}(\boldsymbol{y}^\top, 1)\begin{pmatrix} \widetilde{\boldsymbol{H}}(t), & -\widetilde{\boldsymbol{g}}(t) \\ -\widetilde{\boldsymbol{g}}(t)^\top, & -2\widetilde{f}(t) \end{pmatrix}\begin{pmatrix} \boldsymbol{y} \\ 1 \end{pmatrix}.$$

In addition, the following inequality holds:

$$CE(t, \boldsymbol{y}, w) \leq CE^{no}(t, \boldsymbol{y}, w).$$

We can also connect the certain equivalent of a Γ-constrained portfolio with the Γ-neutral certainty equivalent.

Theorem 6.15. *The certainty equivalent admits the following decomposition:*

$$CE(t, \boldsymbol{y}, w) = \underbrace{CE_0(t, \boldsymbol{y}, w)}_{\boldsymbol{\Gamma}\text{-neutral } CE} \underbrace{- \frac{1}{\gamma}\left(\boldsymbol{y}^\top \boldsymbol{\Psi}(t)\boldsymbol{c} + \boldsymbol{\beta}(t)\boldsymbol{c} + \boldsymbol{c}^\top \boldsymbol{\Lambda}(t)\boldsymbol{c}\right)}_{\text{opportunity multiplier by } \boldsymbol{c}},$$

as indicated in the equation, the first part in the certainty equivalent corresponds to $\boldsymbol{\Gamma}$*-neutral case, and the second part is a multiplier in quadratic form w.r.t* $\mathbf{c}$.

Proof. Substituting in the decomposition given in Lemma 6.9 for $\boldsymbol{g}$ and f, we obtain the decomposition for certainty equivalents as above. □

By Theorem 6.15, the second part is exponential of a quadratic form, whose convexity is guaranteed by positive semidefinite property of $\boldsymbol{\Lambda}(t)$. Therefore, the corresponding certainty equivalent $CE(t, \boldsymbol{y}, w)$ has a global maximum when optimized over the constraint vector $\boldsymbol{c}$. This leads us to define the optimal constraint vector $\boldsymbol{c}^*$.

Definition 6.16. The $\boldsymbol{c}$ for maximizing the certainty equivalent in exponential utility is denoted by

$$\boldsymbol{c}^*(t, \boldsymbol{y}, w) = \arg\max_{\boldsymbol{c} \in \mathbb{R}^d} CE(t, \boldsymbol{y}, w),$$

moreover, if $\boldsymbol{\Lambda}(t)$ is strictly positive, then $\boldsymbol{c}^*(t, \boldsymbol{y}, w)$ is unique.

With Theorem 6.15, seeking the optimal $\boldsymbol{c}$ is equivalent to maximize the part with the opportunity multiplier generated by $\boldsymbol{c}$. And if $\boldsymbol{\Lambda}(t)$ is strictly positive, then $\boldsymbol{c}^*$ admits the formula

$$\boldsymbol{c}^*(t, \boldsymbol{y}, w) = -\boldsymbol{\Lambda}(t)^{-1}\left(\Psi(t)^\top \boldsymbol{y} + \boldsymbol{\beta}(t)^\top\right).$$

Remark 6.17.

(1) Like Γ, the constraint vector $\boldsymbol{c}$ is exogenously imposed on the portfolio, which means it is not a choice by the investor. The vector $\boldsymbol{c}^*$ serves as reference that allows the investor to observe the difference between $\boldsymbol{c}$ and $\boldsymbol{c}^*$ and compare the corresponding certainty equivalents. In other words, the investor can determine the value lost, measured by the reduction in certainty equivalent, due to using the constraint vector $\boldsymbol{c}$ instead of $\boldsymbol{c}^*$.

(2) Note that $\boldsymbol{c}^*$ does not depend on wealth w. Moreover, it has linear dependence on the basis $\boldsymbol{y}$.

(3) If $\boldsymbol{\Gamma} = \mathbf{1}_{N\times 1}$, $\boldsymbol{c} \equiv c$ (a scalar) represents the required net exposure of portfolio, with 1 meaning 100% exposure.

We provide some numerical examples for certainty equivalents in Section 6.4.

6.4 Numerical Illustration

In this section, we simulate the asset's spot prices, futures prices and optimal strategies for our basis model. We also generate empirical wealth distribution at terminal time and certainty equivalents for different constraints and parameters. Primarily, we consider a market with two different assets S_1 and S_2 and three futures that $F_{1,1}$ is written on S_1, while $F_{2,1}$ and $F_{2,2}$ are written on S_2. Their maturities are $T_{1,1} = T_{2,1} = 2/12$ year and $T_{2,2} = 3/12$ year, respectively. Then, our trading horizon is set to be $T = 1/12$ year, strictly less than the futures maturities. We use 'months' or 'trading day' as the w axis in figures. For clarification, we assume there be 252 trading days in a year and 21 trading days for a month. Therefore, our trading horizon is 21 trading days in total.

For other model parameters, we list them as follows.

$$\boldsymbol{\mu} = (0.1, 0.2)^\top, \quad \boldsymbol{m} = (0,0,0)^\top, \quad (\kappa_{1,1}, \kappa_{2,1}, \kappa_{2,2}) = (0.5, 0.5, 0.5),$$

$$\widetilde{\boldsymbol{\Sigma}}_{\boldsymbol{S}} = \begin{bmatrix} 0.5 & 0 \\ 0.3 & 0.4 \end{bmatrix}, \quad \widetilde{\boldsymbol{\Sigma}}_{\boldsymbol{YS}} = \begin{bmatrix} -0.25 & 0 \\ -0.15 & -0.2 \\ -0.15 & -0.2 \end{bmatrix}, \quad \widetilde{\boldsymbol{\Sigma}}_{\boldsymbol{Y}} = \begin{bmatrix} 0.1 & 0 & 0 \\ 0 & 0.1 & 0 \\ 0 & 0 & 0.1 \end{bmatrix}.$$

In turn, we obtain the values for $\boldsymbol{\theta}$, $\widetilde{\boldsymbol{\Sigma}}_{\boldsymbol{FS}}$ and $\widetilde{\boldsymbol{\Sigma}}_{\boldsymbol{F}}$ by applying equations (6.2), (6.5) and (6.6):

$$\boldsymbol{\theta} = \begin{bmatrix} 0.18625 \\ 0.27625 \\ 0.23625 \end{bmatrix}, \quad \widetilde{\boldsymbol{\Sigma}}_{\boldsymbol{FS}} = \begin{bmatrix} 0.25 & 0 \\ 0.15 & 0.2 \\ 0.15 & 0.2 \end{bmatrix}, \quad \widetilde{\boldsymbol{\Sigma}}_{\boldsymbol{F}} = \begin{bmatrix} 0.1 & 0 & 0 \\ 0 & 0.1 & 0 \\ 0 & 0 & 0.1 \end{bmatrix}.$$

In Figure 6.1, we simulate the sample paths for the underling asset price $\boldsymbol{S}_t$, futures price $\boldsymbol{F}_t$ and log-basis $\boldsymbol{Y}_t$. Figure 6.1(a) shows the price paths for asset S_1 and its 2-month futures $F_{1,1}$. The price paths for asset S_2 and its two futures $F_{2,1}$ and $F_{2,2}$ are presented in Figure 6.1(b). Figure 6.1(c) shows the simulated paths for the three log-bases associated with the three futures $F_{1,1}$, $F_{2,1}$, and $F_{2,2}$. The initial prices for both assets are \$10. The log-basis vector $\boldsymbol{Y}_t$ is a multidimensional Brownian bridge that starts

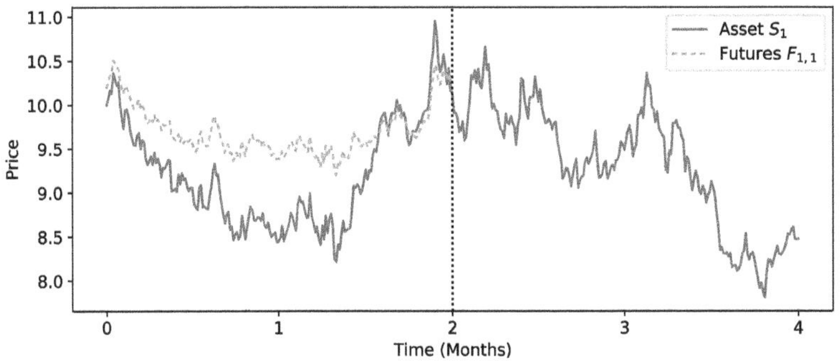

(a)

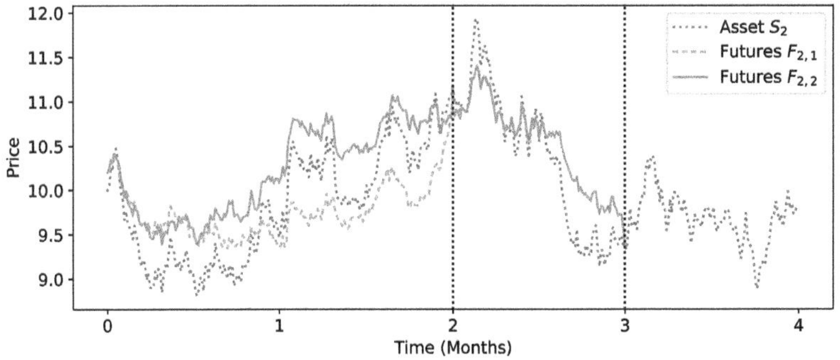

(b)

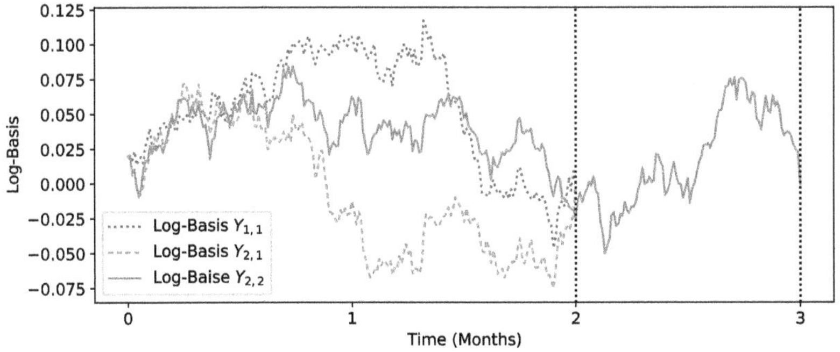

(c)

Fig. 6.1. Simulated path for assets prices $\boldsymbol{S}_t$, futures prices $\boldsymbol{F}_t$ and log-bases $\mathbf{Y}_t$. (a) asset S_1 with its 2-month futures $F_{1,1}$. (b) asset S_2 with its 2-month futures $F_{2,1}$ and 3-month futures $F_{2,2}$. (c) log-basis $\mathbf{Y}_t$. Initial value: $\boldsymbol{S}_0 = (10, 10)^\top$ and $\mathbf{Y}_0 = (0.02, 0.02, 0.02)^\top$.

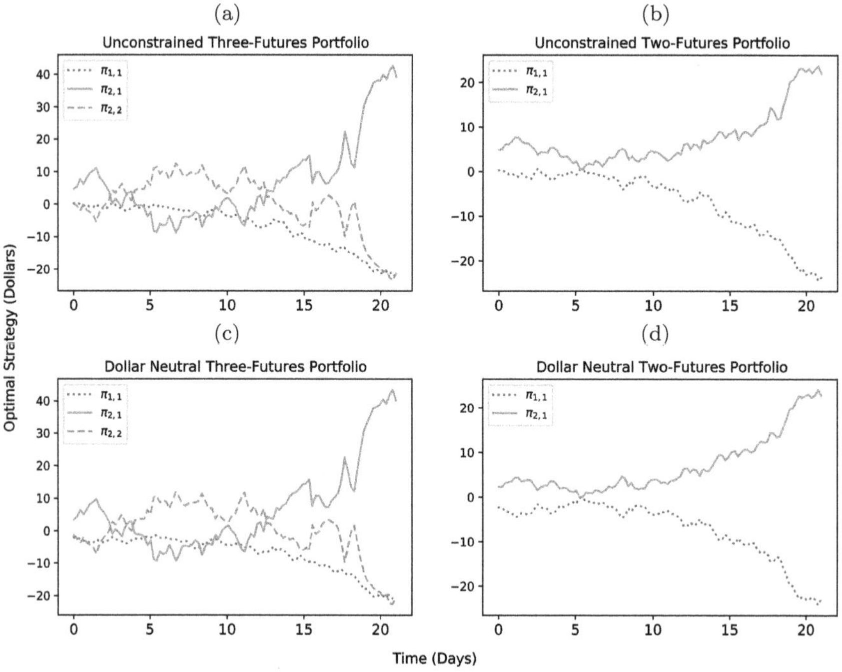

Fig. 6.2. Optimal strategies. (a) Unconstrained three-futures portfolio. (b) Unconstrained two-futures portfolio. (c) Dollar neutral three-futures portfolio. (d) Dollar neutral two futures portfolio.

from $\boldsymbol{Y}_0 = (0.02, 0.02, 0.02)^\top$ and converges to zero, which guarantees that each futures price is equal to corresponding asset price at maturity, which are $T_{1,1} = T_{2,1} = 2$ months and $T_{2,2} = 3$ months, respectively.

In Figure 6.2, we illustrate the path behaviors of the optimal strategies under different settings. Figures 6.2(a) and 6.2(c) plots, we illustrate optimal unconstrained strategy and optimal dollar neutral strategy for three-futures portfolio. For the dollar neutral strategy, the sum of positions for all futures must be equal to zero in order to satisfy the constraint. Interestinly, the two sets of strategies look similar, which means that the unconstrained optimal strategy is close to being dollar neutral. Also, we observe that in both cases the portfolio has a long-short combinations with opposite positions in two sets of futures, i.e. $F_{2,1}$ vs $F_{1,1}$ and $F_{2,2}$. The same phenomenon can be observed for the two-futures portfolios (right panel) as well. With two futures $F_{1,1}$ and $F_{2,1}$, the investor tends to long one and short the other whether the dollar neutral constraint is enforced or not. Lastly, in our model the optimal strategies depend crucially on the log-basis vector process $\boldsymbol{Y}_t$ over time. This is unlike the optimal strategies obtained by Leung and Yan (2018, 2019), which are time-deterministic.

In Figure 6.3, we present the empirical distributions of terminal wealth associated with portfolios with different number of futures and constraints under different risk aversion parameters γ. From Figures 6.3(a)–6.3(d), the wealth distributions are for (i) three-futures portfolio with $\gamma = 0.5$; (ii) two-futures portfolio with $\gamma = 0.5$; (iii) three-futures portfolio with $\gamma = 0.1$; (iv) two-futures portfolio with $\gamma = 0.1$. Like Figure 6.2, we use the combination of $F_{1,1}$ and $F_{2,1}$ as the two-futures portfolio. We note that for the two-futures portfolio, the market neutral constraint matrix

$$\begin{aligned}\boldsymbol{\Gamma} &= \widetilde{\Sigma}_{\boldsymbol{FS}}[1,2;1,2]\widetilde{\Sigma}_{\boldsymbol{S}}^{\top} \\ &= \begin{bmatrix} 0.25 & 0 \\ 0.15 & 0.2 \end{bmatrix} \begin{bmatrix} 0.5 & 0.3 \\ 0 & 0.4 \end{bmatrix},\end{aligned}$$

is invertible. This implies $\boldsymbol{\pi}^* = \mathbf{0}$, meaning that the only feasible strategy is not to invest at all. Therefore, the market neutral strategy does not appear in the second and fourth charts.

Among the strategies, the market neutral strategy leads to a more narrow and centralized distribution of terminal wealth, while the unconstrained portfolio wealth has a wider distribution with the heaviest tails. This reflects that the unconstrained strategy has the highest degree of freedom and the investor ends up bearing more risk when no constraint is imposed. On the other hand, risk aversion also plays a role in the terminal wealth distribution. If we compare Figures 6.3(a) and 6.3(b) to Figures 6.3(c) and 6.3(d), the more risk averse investor ($\gamma = 0.5$), the distribution is more centralized than those for the less risk averse investor ($\gamma = 0.1$). Given that the optimal strategy π^* is inversely proportional to γ according to Theorem 6.3, the less risk averse investor is likely to hold larger net positions in futures and bear more risk as a result.

For the wealth distributions in Figure 6.3, we provide the summary statistics, including average one-month P&L, standard deviation, and quantiles, in Table 6.1. For each portfolio, we define

$$\text{Average Net Exposure} = \frac{1}{T}\int_0^T \mathbf{1}^\top \boldsymbol{\pi}_t dt,$$

$$\text{Maximum Net Exposure} = \max_{0\le t\le T} |\mathbf{1}^\top \boldsymbol{\pi}_t|,$$

which are reported in the table.

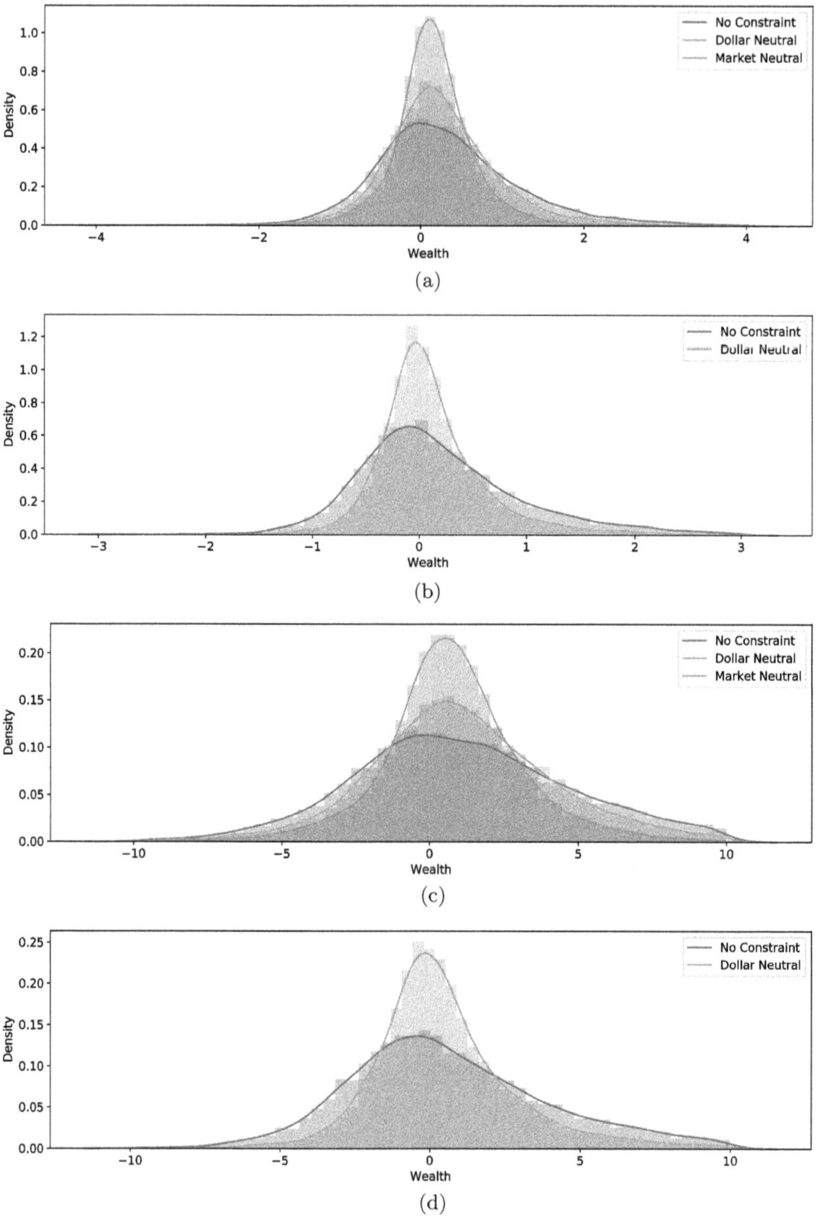

Fig. 6.3. The distributions of terminal wealth for the unconstrained and constrained portfolios. From (a)–(d): (i) three-futures portfolio with $\gamma = 0.5$; (ii) two-futures portfolio with $\gamma = 0.5$; (iii) three-futures portfolio with $\gamma = 0.1$; (iv) two-futures portfolio with $\gamma = 0.1$.

Table 6.1. Average one-month P&L, standard deviation and quantiles for the wealth distributions in Figure 6.3.

	3 Futures			2 Futures	
$\gamma = 0.1$	NC	DN	MN	NC	DN
Average one-month P&L	1.68	1.27	0.77	1.10	0.58
Standard deviation	4.67	3.50	2.52	4.18	2.61
Lower quartile	−1.27	−0.84	−0.51	−1.56	−0.89
Median	1.11	0.96	0.69	0.35	0.18
Upper quartile	3.97	3.07	2.02	2.97	1.57
Average net exposure	27.06	0	0	26.31	0
Maximum net exposure	61.11	0	0	66.92	0
	3 Futures			2 Futures	
$\gamma = 0.5$	NC	DN	MN	NC	DN
Average one-month P&L	0.34	0.25	0.15	0.22	0.12
Standard deviation	0.93	0.70	0.50	0.84	0.52
Lower quartile	−0.25	−0.17	−0.10	−0.31	−0.18
Median	0.22	0.19	0.14	0.07	0.04
Upper quartile	0.79	0.61	0.40	0.59	0.31
Average net exposure	5.41	0	0	5.26	0
Maximum net exposure	12.22	0	0	13.38	0

Note: NC, DN, and MN stand for "no constraint", "dollar neutral" and "market neutral", respectively.

The statistics for the less risk averse investor ($\gamma = 0.1$) are summarized in the upper half of Table 6.1. From the average one-month P&L, we see that the unconstrained strategy is most profitable and the market neutral strategy is the least when trading three futures in the portfolio with $\gamma = 0.1$. Also, in terms of standard deviation, the unconstrained portfolio is the riskiest and the market neutral strategy is the most conservative, which is consistent with Figure 6.3.

Comparing the top and bottom parts of the table, we observe that the average one-month P&L, standard deviation and net exposure are inversely proportional to risk aversion for all three strategies. For example, the less risk-averse investor (with $\gamma = 0.1$) has a higher average P&L but also higher standard deviation than the more risk-averse investor (with $\gamma = 0.5$). This phenomenon is expected from the optimal strategy formulas in (6.20) and (6.31).

In Figure 6.4, we plot the certainty equivalents for the dollar constraint portfolios, where the total amount invested $\mathbf{1}^\top \boldsymbol{\pi}$ is set to be a fixed

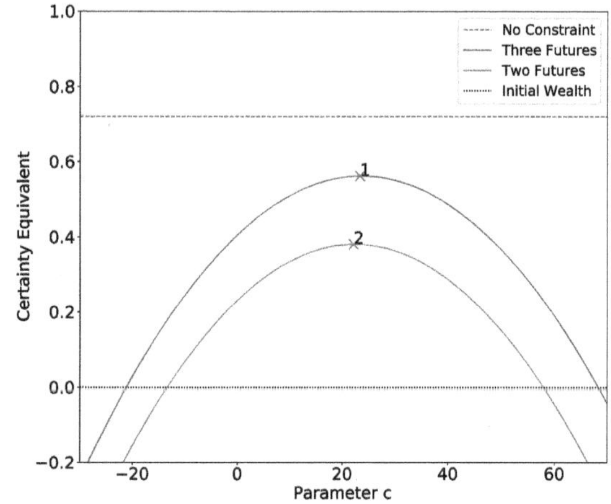

(a)

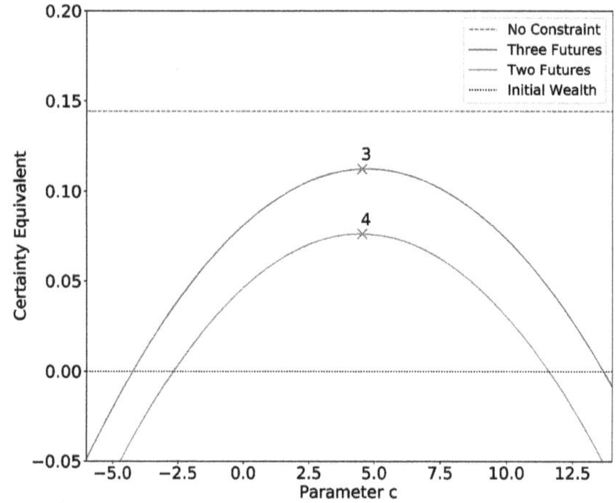

(b)

Fig. 6.4. The certainty equivalent (CE) as the function of constraint parameter c. (a) and (b) plots correspond to the cases with $\gamma = 0.1$ and $\gamma = 0.5$ respectively. In each plot, the dashed line shows the CE for unconstrained three-futures portfolio, which is independent of c. The top curve (marked with number 1 or 3) represents the CE for the constrained three-futures portfolio, and the bottom curve (marked with number 2 or 4) represents the CE for the constrained two-futures portfolio. The black dashed line represents the initial wealth. Each cross marks the optimal parameter c^* (on the x-axis) and the corresponding maximum certainty equivalent $CE*$ (on the y-axis). Optimal parameters: $c_1^* = 23.4$, $c_2^* = 22.2$, $c_3^* = 4.54$ and $c_4^* = 4.53$, and maximum certainty equivalents: $CE_1^* = 0.562$, $CE_2^* = 0.381$, $CE_3^* = 0.112$ and $CE_4^* = 0.076$.

parameter c. In other words, the dollar neutral constraint is a special case with $c = 0$.

The top and bottom curves represent the certainty equivalents for the three-futures portfolio and two-futures portfolio, respectively. Since, with respect to dollar constraint, the admissible strategy for two-futures portfolio is always the admissible strategy for three-futures portfolio, the three-futures portfolio's certainty equivalent must be higher than the two-futures portfolio's certainty equivalent for any c.

In fact, Figure 6.4 confirms this. For comparison, we use a dashed line to mark the certainty equivalent for the no constraint three-futures portfolio. As expected, it is higher than the certainty equivalent for the constrained portfolios. We also put a dark dashed line to show the initial wealth. If portfolio's certainty equivalent is lower than its initial wealth, then it is not worthwhile to trade.

For each portfolio, there is an optimal constraint parameter c^* (marked by a cross) that maximizes the certainty equivalent. For the less risk averse investor ($\gamma = 0.1$), the best parameters c^* for the three-futures and two-futures portfolios are $c_1^* = 23.4$ and $c_2^* = 22.2$, and the corresponding certainty equivalents are $CE_1^* = 0.562$ and $CE_2^* = 0.381$, respectively.

In contrast, for the more risk averse investor ($\gamma = 0.5$), the highest certainty equivalents for two portfolios are lower with $CE_3^* = 0.112$ and $CE_4^* = 0.076$, and the corresponding constraint parameters are $c_3^* = 4.54$ and $c_4^* = 4.53$, respectively. Generally, the less risk averse investor can achieve a higher certainty equivalent than the more risk averse investor for each portfolio.

Lastly, we present in Figure 6.5 the certainty equivalent as a function of the constraint vector $\boldsymbol{c} = (c_1, c_2)$ for the market-constrained three-futures portfolios. For the market constraint, the strategy satisfies $(\widetilde{\boldsymbol{\Sigma}}_{\boldsymbol{FS}}\widetilde{\boldsymbol{\Sigma}}_{\boldsymbol{S}}^\top)^\top \boldsymbol{\pi} = \boldsymbol{c}$ for some fixed $\boldsymbol{c} \in \mathbb{R}^{M\times 1}$, where $M = 2$ is the number of assets in our example. The dashed contours denote the certainty equivalent of value zero, same as the initial wealth $W_0 = 0$. For portfolios corresponding to the interior of the contour, the certainty equivalent is higher than the investor's initial wealth, which means it is worthwhile to trade. The contour region for less risk averse investor ($\gamma = 0.1$) is bigger than the contour region for more risk averse investor ($\gamma = 0.5$). Moreover, the less risk averse investor achieves a higher certainty equivalent with same portfolio. In both plots, the optimal constraint parameters $\boldsymbol{c}^*$ are marked by crosses, such that $\boldsymbol{c}_1^* = (1.71, 3.01)^\top$ and the corresponding certainty equivalent $CE_1^* = 0.39$ for $\gamma = 0.1$, and $\boldsymbol{c}_2^* = (0.34, 0.59)^\top$ and $CE_2^* = 0.08$ for $\gamma = 0.5$.

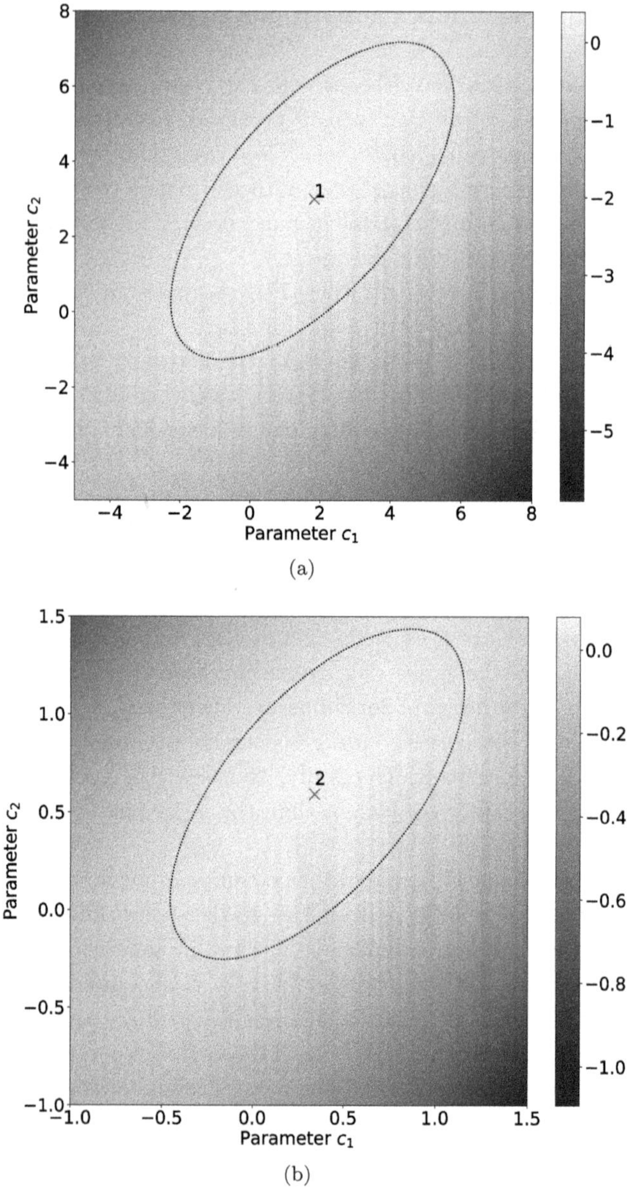

Fig. 6.5. The certainty equivalent (CE) for the market-constrained three-futures portfolio with different risk aversion levels ((a) $\gamma = 0.1$; (b) $\gamma = 0.5$). The dashed contours denote the certainty equivalent with initial wealth $W_0 = 0$, and the optimal parameters $\boldsymbol{c}^*$ are marked by crosses. The optimized constraint vectors are $\mathbf{c}_1^* = (1.71, 3.01)^\top$ (a) and $\mathbf{c}_2^* = (0.34, 0.59)^\top$ (b), and the corresponding certainty equivalents are $CE_1^* = 0.39$ and $CE_2^* = 0.08$ respectively.

6.5 Further Applications

We have analyzed constrained dynamic futures portfolios in continuous time under a multidimensional stochastic basis model. One key element of our utility maximization problem is the incorporation of general portfolio constraints on the futures positions, which captures the dollar neutral and market neutral constraints. We have derived the optimal strategies in the unconstrained and constrained cases by solving the associated HJB equations, and presented a number of decomposition results for the corresponding optimal strategies and certainty equivalents.

Apart from basis trading, the multidimensional stochastic basis model should also be useful for quantitative risk management. The joint dynamics of spot prices and their bases can be applied to risk estimation.

6.6 Proofs

In this section, we provide the detailed proofs for the theorems and propositions presented in this chapter.

Proof of Theorem 6.2

Before proving the Theorem 6.2, we begin with a lemma.

Lemma 6.18. *$\boldsymbol{\Sigma}^{no}$ is positive semidefinite.*

Proof. By inspecting the structure of $\boldsymbol{\Sigma}^{no}$ in (6.15), it suffices to show that $\boldsymbol{\Sigma_Y} - \boldsymbol{\Sigma}_{\boldsymbol{FY}}^\top \boldsymbol{\Sigma}_{\boldsymbol{F}}^{-1} \boldsymbol{\Sigma}_{\boldsymbol{FY}}$ is positive semidefinite.

Define

$$\boldsymbol{A_S} = \widetilde{\boldsymbol{\Sigma}}_{\boldsymbol{FS}}^\top - \widetilde{\boldsymbol{\Sigma}}_{\boldsymbol{YS}}^\top.$$

Then by (6.10), (6.11), (6.12) and the fact that $\widetilde{\boldsymbol{\Sigma}}_{\boldsymbol{F}} = \widetilde{\boldsymbol{\Sigma}}_{\boldsymbol{Y}}$, we have

$$\boldsymbol{\Sigma}_{\boldsymbol{FY}}^\top \boldsymbol{\Sigma}_{\boldsymbol{F}}^{-1} \boldsymbol{\Sigma}_{\boldsymbol{FY}} = (\boldsymbol{\Sigma_F} - \widetilde{\boldsymbol{\Sigma}}_{\boldsymbol{FS}} \boldsymbol{A_S})^\top \boldsymbol{\Sigma}_{\boldsymbol{F}}^{-1} (\boldsymbol{\Sigma_F} - \widetilde{\boldsymbol{\Sigma}}_{\boldsymbol{FS}} \boldsymbol{A_S}).$$

This in turns gives the equality:

$$\boldsymbol{\Sigma_Y} - \boldsymbol{\Sigma}_{\boldsymbol{FY}}^\top \boldsymbol{\Sigma}_{\boldsymbol{F}}^{-1} \boldsymbol{\Sigma}_{\boldsymbol{FY}} = \boldsymbol{A}_{\boldsymbol{S}}^\top (\boldsymbol{I} - \widetilde{\boldsymbol{\Sigma}}_{\boldsymbol{FS}}^\top \boldsymbol{\Sigma}_{\boldsymbol{F}}^{-1} \widetilde{\boldsymbol{\Sigma}}_{\boldsymbol{FS}}) \boldsymbol{A_S}.$$

Hence, we only need to show $\boldsymbol{I} - \widetilde{\boldsymbol{\Sigma}}_{\boldsymbol{FS}}^\top \boldsymbol{\Sigma}_{\boldsymbol{F}}^{-1} \widetilde{\boldsymbol{\Sigma}}_{\boldsymbol{FS}}$ is the non-negative matrix. Therefore, it remains to show that the eigenvalues of $\widetilde{\boldsymbol{\Sigma}}_{\boldsymbol{FS}}^\top \boldsymbol{\Sigma}_{\boldsymbol{F}}^{-1} \widetilde{\boldsymbol{\Sigma}}_{\boldsymbol{FS}}$ are bounded between $[-1, 1]$.

Now, suppose $\boldsymbol{v}$ is an eigenvector for $\widetilde{\boldsymbol{\Sigma}}_{\boldsymbol{FS}}^\top \boldsymbol{\Sigma}_{\boldsymbol{F}}^{-1} \widetilde{\boldsymbol{\Sigma}}_{\boldsymbol{FS}}$ such that

$$\widetilde{\boldsymbol{\Sigma}}_{\boldsymbol{FS}}^\top \boldsymbol{\Sigma}_{\boldsymbol{F}}^{-1} \widetilde{\boldsymbol{\Sigma}}_{\boldsymbol{FS}} \boldsymbol{v} = \lambda \boldsymbol{v}.$$

With this, we obtain

$$\begin{aligned}
\lambda \boldsymbol{v}^\top \boldsymbol{v} &= \boldsymbol{v}^\top \widetilde{\boldsymbol{\Sigma}}_{\boldsymbol{FS}}^\top \boldsymbol{\Sigma}_{\boldsymbol{F}}^{-1} \widetilde{\boldsymbol{\Sigma}}_{\boldsymbol{FS}} \boldsymbol{v} \\
&= \boldsymbol{v}^\top \widetilde{\boldsymbol{\Sigma}}_{\boldsymbol{FS}}^\top \boldsymbol{\Sigma}_{\boldsymbol{F}}^{-1} \boldsymbol{\Sigma}_{\boldsymbol{F}} \boldsymbol{\Sigma}_{\boldsymbol{F}}^{-1} \widetilde{\boldsymbol{\Sigma}}_{\boldsymbol{FS}} \boldsymbol{v} \\
&\geq \boldsymbol{v}^\top \widetilde{\boldsymbol{\Sigma}}_{\boldsymbol{FS}}^\top \boldsymbol{\Sigma}_{\boldsymbol{F}}^{-1} \widetilde{\boldsymbol{\Sigma}}_{\boldsymbol{FS}} \widetilde{\boldsymbol{\Sigma}}_{\boldsymbol{FS}}^\top \boldsymbol{\Sigma}_{\boldsymbol{F}}^{-1} \widetilde{\boldsymbol{\Sigma}}_{\boldsymbol{FS}} \boldsymbol{v} \\
&= \lambda^2 \boldsymbol{v}^\top \boldsymbol{v}.
\end{aligned}$$

From this inequality, we conclude that $\lambda \in [0, 1]$. □

Now, we give the proof for Theorem 6.2 on the matrix Riccati equation and the associated expression for the optimal strategy. According to Appendix A in Angoshtari and Leung (2020), the Riccati equation (6.16) has a unique symmetric non-negative definite solution due to the fact that $\boldsymbol{\Sigma}_{\boldsymbol{F}}^{-1}$ and $\boldsymbol{\Sigma}^{no}$ are positive semidefinite.

Next, performing optimization in (6.14) and assuming that $\partial_{ww} u \leq 0$ (which will be verified later), we obtain the first-order condition for the optimal strategy $\boldsymbol{\pi}_t^*$,

$$\boldsymbol{a}^{no}(t, \boldsymbol{y}, w) + (\partial_{ww} u) \boldsymbol{\Sigma}_{\boldsymbol{F}} \boldsymbol{\pi}_t^* = 0.$$

Then, we can write the optimal strategy as

$$\boldsymbol{\pi}_t^* = -\frac{\boldsymbol{\Sigma}_{\boldsymbol{F}}^{-1} \boldsymbol{a}^{no}(t, \boldsymbol{y}, w)}{\partial_{ww} u}. \tag{6.34}$$

Plugging (6.34) back to the HJB (6.14), we have

$$\partial_t u^{no} + \mathcal{L} u^{no} - \frac{1}{2 \partial_{ww} u^{no}} \boldsymbol{a}^{no}(t, \boldsymbol{y}, w)^\top \boldsymbol{\Sigma}_{\boldsymbol{F}}^{-1} \boldsymbol{a}^{no}(t, \boldsymbol{y}, w) = 0.$$

With the transformation

$$u^{no}(t, \boldsymbol{y}, w) = U(w) \exp\left(-\frac{1}{2} \boldsymbol{y}^\top \boldsymbol{H}^{no}(t) \boldsymbol{y} + \boldsymbol{y}^\top \boldsymbol{g}^{no}(t) + f^{no}(t) \right),$$

where $\boldsymbol{H}^{no}(t) \in \mathbb{R}^{N \times N}$ is a symmetric matrix and $\boldsymbol{g}^{no}(t) \in \mathbb{R}^N$, we obtain the matrix Riccati equation (6.16) for $\boldsymbol{H}^{no}(t)$, ODE system (6.18) for $\boldsymbol{g}^{no}(t)$ and ODE (6.19) for $f^{no}(t)$. In addition, we confirm that

$$\partial_{ww} u = \gamma^2 u \leq 0.$$

Hence, the assertions in Theorem 6.2 have been shown.

Proof of Theorem 6.3

Before proving Theorem 6.3, we first give two lemmas.

Lemma 6.19. *The matrix* $\boldsymbol{\Sigma}_{\boldsymbol{F}}^{-1} - \boldsymbol{\Sigma}_{\boldsymbol{\Gamma}}$ *is positive semidefinite.*

Proof. By (6.25), we have

$$\boldsymbol{\Sigma}_{\boldsymbol{F}}^{-1} - \boldsymbol{\Sigma}_{\boldsymbol{\Gamma}} = \boldsymbol{\Sigma}_{\boldsymbol{F}}^{-1} - \boldsymbol{\Sigma}_{\boldsymbol{F}}^{-1}\boldsymbol{\Gamma}\boldsymbol{D}_{\boldsymbol{\Gamma}}^{-1}\boldsymbol{\Gamma}^{\top}\boldsymbol{\Sigma}_{\boldsymbol{F}}^{-1}.$$

Since $\boldsymbol{\Sigma}_{\boldsymbol{F}}$ is symmetric positive definite matrix, we define an auxiliary matrix

$$\boldsymbol{A} = \sqrt{\boldsymbol{\Sigma}_{\boldsymbol{F}}^{-1}}\boldsymbol{\Gamma} \in \mathbb{R}^{N\times d}.$$

Therefore, we can write

$$\boldsymbol{D}_{\boldsymbol{\Gamma}} = \boldsymbol{A}^{\top}\boldsymbol{A}.$$

Then, it suffices to show

$$\sqrt{\boldsymbol{\Sigma}_{\boldsymbol{F}}}(\boldsymbol{\Sigma}_{\boldsymbol{F}}^{-1} - \boldsymbol{\Sigma}_{\boldsymbol{F}}^{-1}\boldsymbol{\Gamma}\boldsymbol{D}_{\boldsymbol{\Gamma}}^{-1}\boldsymbol{\Gamma}^{\top}\boldsymbol{\Sigma}_{\boldsymbol{F}}^{-1})\sqrt{\boldsymbol{\Sigma}_{\boldsymbol{F}}} = \boldsymbol{I} - \boldsymbol{A}(\boldsymbol{A}^{\top}\boldsymbol{A})^{-1}\boldsymbol{A}^{\top},$$

is non-negative definite. Since, $\boldsymbol{A}\boldsymbol{v}$ is the eigenvector of $\boldsymbol{A}(\boldsymbol{A}^{\top}\boldsymbol{A})^{-1}\boldsymbol{A}^{\top}$, for any $\boldsymbol{v} \in \mathbb{R}^d$ and $\text{rank}(\boldsymbol{A}) = d$, the eigenvalues for $\boldsymbol{A}(\boldsymbol{A}^{\top}\boldsymbol{A})^{-1}\boldsymbol{A}^{\top}$ are 0 or 1. Therefore, $\boldsymbol{I} - \boldsymbol{A}(\boldsymbol{A}^{\top}\boldsymbol{A})^{-1}\boldsymbol{A}^{\top}$ is positive semidefinite. □

Lemma 6.20. *The matrix* $\boldsymbol{\Sigma}$ *is positive semidefinite.*

Proof. By (6.26), we can decompose $\boldsymbol{\Sigma}$ into two components, i.e.

$$\boldsymbol{\Sigma} = \boldsymbol{\Sigma}^{no} + \boldsymbol{\Sigma}_{\boldsymbol{FY}}^{\top}\boldsymbol{\Sigma}_{\boldsymbol{\Gamma}}\boldsymbol{\Sigma}_{\boldsymbol{FY}}.$$

According to Lemma 6.18, $\boldsymbol{\Sigma}^{no} \geq 0$. Besides, the second term is also positive semidefinite. Therefore, $\boldsymbol{\Sigma}$ is positive semidefinite. □

Now, we show the proof for Theorem 6.3. According to Appendix A in Angoshtari and Leung (2020). To show that the Riccati equation (6.27) has a unique symmetric positive semidefinite solution, it suffices to demonstrate that $\boldsymbol{\Sigma}_{\boldsymbol{F}}^{-1} - \boldsymbol{\Sigma}_{\boldsymbol{\Gamma}}$ and $\boldsymbol{\Sigma}$ being the coefficients of quadratic and constant terms are positive semidefinite, which are proved in Lemmas 6.19 and 6.20.

Next, we turn to solve the HJB equation (6.23) by the method of Lagrange multiplier, similarly seen in Li and Papanicolaou (2019). Let use define

$$\boldsymbol{\lambda}(t) = (\lambda_1(t), \ldots, \lambda_d(t))^\top$$

to be the (vector) Lagrange multiplier, and we define the Lagrangian function for the constrained control model

$$L(t, \boldsymbol{\pi}, \boldsymbol{\lambda}) = \boldsymbol{\pi}^\top \boldsymbol{a}(t, \boldsymbol{y}, w) + \frac{\partial_{ww} u}{2} \boldsymbol{\pi}^\top \boldsymbol{\Sigma}_{\boldsymbol{F}} \boldsymbol{\pi} - \boldsymbol{\lambda}(t)^\top (\boldsymbol{\Gamma}^\top \boldsymbol{\pi} - \boldsymbol{c}). \tag{6.35}$$

Then, it suffices to solve

$$\begin{cases} \nabla_{\boldsymbol{\pi}} L = \boldsymbol{a}(t, \boldsymbol{y}, w) + (\partial_{ww} u) \boldsymbol{\Sigma}_{\boldsymbol{F}} \boldsymbol{\pi} - \boldsymbol{\Gamma} \boldsymbol{\lambda}(t) = 0, \\ \nabla_{\boldsymbol{\lambda}} L = \boldsymbol{\Gamma}^\top \boldsymbol{\pi} - \boldsymbol{c} = 0. \end{cases}$$

Therefore, we obtain

$$\boldsymbol{\pi}_t = -\boldsymbol{\Sigma}_{\boldsymbol{F}}^{-1} \frac{\boldsymbol{a}(t, \boldsymbol{y}, w) - \boldsymbol{\Gamma} \boldsymbol{\lambda}(t)}{\partial_{ww} u}, \tag{6.36}$$

and

$$\boldsymbol{\lambda}(t) = (\boldsymbol{\Gamma}^\top \boldsymbol{\Sigma}_{\boldsymbol{F}}^{-1} \boldsymbol{\Gamma})^{-1} (\boldsymbol{\Gamma}^\top \boldsymbol{\Sigma}_{\boldsymbol{F}}^{-1} \boldsymbol{a}(t, \boldsymbol{y}, w) + (\partial_{ww} u) \boldsymbol{c}).$$

Inserting $\boldsymbol{\lambda}$ back to formula (6.36), we have

$$\boldsymbol{\pi}_t = \boldsymbol{\Sigma}_{\boldsymbol{F}}^{-1} \boldsymbol{\Gamma} \boldsymbol{D}_{\boldsymbol{\Gamma}}^{-1} \boldsymbol{c} - \frac{(\boldsymbol{\Sigma}_{\boldsymbol{F}}^{-1} - \boldsymbol{\Sigma}_{\boldsymbol{\Gamma}}) \boldsymbol{a}(t, \boldsymbol{y}, w)}{\partial_{ww} u}, \tag{6.37}$$

where $\boldsymbol{D}_{\boldsymbol{\Gamma}}$ and $\boldsymbol{\Sigma}_{\boldsymbol{\Gamma}}$ are given by (6.24) and (6.25), respectively.

We verify that the optimal strategy must satisfy the constraints

$$\boldsymbol{\Gamma}^\top \boldsymbol{\pi}_t = \boldsymbol{\Gamma}^\top \boldsymbol{\Sigma}_{\boldsymbol{F}}^{-1} \boldsymbol{\Gamma} \boldsymbol{D}_{\boldsymbol{\Gamma}}^{-1} \boldsymbol{c} - \frac{\boldsymbol{\Gamma}^\top (\boldsymbol{\Sigma}_{\boldsymbol{F}}^{-1} - \boldsymbol{\Sigma}_{\boldsymbol{\Gamma}}) \boldsymbol{a}(t, \boldsymbol{y}, w)}{\partial_{ww} u} = \boldsymbol{c}.$$

To that end, putting the candidate strategy $\boldsymbol{\pi}_t$ back to the HJB equation (6.23), we obtain the PDE

$$\partial_t u + \mathcal{L} u - \frac{\boldsymbol{a}(t, \boldsymbol{y}, w)^\top (\boldsymbol{\Sigma}_{\boldsymbol{F}}^{-1} - \boldsymbol{\Sigma}_{\boldsymbol{\Gamma}}) \boldsymbol{a}(t, \boldsymbol{y}, w)}{2 \partial_{ww} u} + \boldsymbol{c}^\top \boldsymbol{D}_{\boldsymbol{\Gamma}}^{-1} \boldsymbol{\Gamma}^\top \boldsymbol{\Sigma}_{\boldsymbol{F}}^{-1} \boldsymbol{a}(t, \boldsymbol{y}, w)$$
$$+ \frac{\partial_{ww} u}{2} \boldsymbol{c}^\top \boldsymbol{D}_{\boldsymbol{\Gamma}}^{-1} \boldsymbol{c} = 0.$$

With the ansatz

$$u(t, \boldsymbol{y}, w) = U(w) \exp\left(-\frac{1}{2}\boldsymbol{y}^\top \boldsymbol{H}(t)\boldsymbol{y} + \boldsymbol{y}^\top \boldsymbol{g}(t) + f(t)\right),$$

where $\boldsymbol{H}(t) \in \mathbb{R}^{N\times N}$ is a symmetric matrix and $\boldsymbol{g}(t) \in \mathbb{R}^N$, we obtain the matrix Riccati equation (6.27) for $\boldsymbol{H}(t)$, ODE system (6.29) for $\boldsymbol{g}(t)$, and ODE (6.30) for $f(t)$. Lastly, putting $u(t, \boldsymbol{y}, w)$ into the (6.37), we obtain the formula of optimal strategy.

Proof of Lemma 6.9

First, a direct observation from matrix Riccati equation (6.27) shows that $\boldsymbol{H}(t)$ doesn't depend on $\boldsymbol{c}$. Then, substituting the decomposition (6.32) into the ODE system (6.29) and ODE (6.30), we obtain the ODE system for $\boldsymbol{g_0}(t) \in \mathbb{R}^N$ as follows:

$$\begin{aligned}
\boldsymbol{g}_0'(t) &= \boldsymbol{K}(t)\boldsymbol{g}_0(t) + \boldsymbol{H}(t)\boldsymbol{m} + \left(\boldsymbol{H}(t)\boldsymbol{\Sigma} + \boldsymbol{K}(t)(\boldsymbol{\Sigma}_{\boldsymbol{F}}^{-1} - \boldsymbol{\Sigma}_{\boldsymbol{\Gamma}})\boldsymbol{\Sigma}_{\boldsymbol{FY}}\right)\boldsymbol{g}_0(t) \\
&\quad + \left(\boldsymbol{K}(t) - \boldsymbol{H}(t)\boldsymbol{\Sigma}_{\boldsymbol{FY}}^\top\right)(\boldsymbol{\Sigma}_{\boldsymbol{F}}^{-1} - \boldsymbol{\Sigma}_{\boldsymbol{\Gamma}})\boldsymbol{\mu}_{\boldsymbol{F}}, \\
\boldsymbol{g}_0(T) &= \boldsymbol{0}_{N\times 1},
\end{aligned}$$

and the ODE for $f_0(t)$:

$$\begin{aligned}
f_0'(t) &= -\boldsymbol{m}^\top \boldsymbol{g}_0(t) + \frac{1}{2} tr\left(\boldsymbol{\Sigma}_{\boldsymbol{Y}}\boldsymbol{H}(t)\right) - \frac{1}{2}\boldsymbol{g}_0(t)^\top \boldsymbol{\Sigma}_{\boldsymbol{Y}}\boldsymbol{g}_0(t) \\
&\quad + \frac{1}{2}\left(\boldsymbol{\theta} + \boldsymbol{\Sigma}_{\boldsymbol{FY}}\boldsymbol{g}_0(t)\right)^\top (\boldsymbol{\Sigma}_{\boldsymbol{F}}^{-1} - \boldsymbol{\Sigma}_{\boldsymbol{\Gamma}})\left(\boldsymbol{\theta} + \boldsymbol{\Sigma}_{\boldsymbol{FY}}\boldsymbol{g}_0(t)\right), \\
f_0(T) &= 0,
\end{aligned}$$

along with the ODEs for $\boldsymbol{\Psi}(t), \boldsymbol{\beta}(t), \boldsymbol{\Lambda}(t)$,

$$\begin{aligned}
\boldsymbol{\Psi}'(t) &= \boldsymbol{K}(t)\boldsymbol{\Psi}(t) + \left(\boldsymbol{H}(t)\boldsymbol{\Sigma} + \boldsymbol{K}(t)(\boldsymbol{\Sigma}_{\boldsymbol{F}}^{-1} - \boldsymbol{\Sigma}_{\boldsymbol{\Gamma}})\boldsymbol{\Sigma}_{\boldsymbol{FY}}\right)\boldsymbol{\Psi}(t) \\
&\quad + \gamma\left(\boldsymbol{K}(t) - \boldsymbol{H}(t)\boldsymbol{\Sigma}_{\boldsymbol{FY}}^\top\right)\boldsymbol{\Sigma}_{\boldsymbol{F}}^{-1}\boldsymbol{\Gamma}\boldsymbol{D}_{\boldsymbol{\Gamma}}^{-1}, \\
\boldsymbol{\beta}'(t) &= -\boldsymbol{m}^\top \boldsymbol{\Psi}(t) - \boldsymbol{g}_0(t)\boldsymbol{\Sigma}\boldsymbol{\Psi}(t) + \boldsymbol{\mu}_{\boldsymbol{F}}^\top(\boldsymbol{\Sigma}_{\boldsymbol{F}}^{-1} - \boldsymbol{\Sigma}_{\boldsymbol{\Gamma}})\boldsymbol{\Sigma}_{\boldsymbol{FY}} \\
&\quad + \gamma(\boldsymbol{\mu}_{\boldsymbol{F}} + \boldsymbol{g}_0(t)\boldsymbol{\Sigma}_{\boldsymbol{FY}})^\top \boldsymbol{\Sigma}_{\boldsymbol{F}}^{-1}\boldsymbol{\Gamma}\boldsymbol{D}_{\boldsymbol{\Gamma}}^\top, \\
\boldsymbol{\Lambda}'(t) &= -\frac{1}{2}\boldsymbol{\Psi}(t)^\top \boldsymbol{\Sigma}\boldsymbol{\Psi}(t) + \frac{\gamma}{2}\boldsymbol{\Psi}(t)^\top \boldsymbol{\Sigma}_{\boldsymbol{FY}}^\top \boldsymbol{\Sigma}_{\boldsymbol{F}}^{-1}\boldsymbol{\Gamma}\boldsymbol{D}_{\boldsymbol{\Gamma}}^{-1} \\
&\quad + \frac{\gamma}{2}\boldsymbol{D}_{\boldsymbol{\Gamma}}^{-1}\boldsymbol{\Gamma}^\top \boldsymbol{\Sigma}_{\boldsymbol{F}}^{-1}\boldsymbol{\Sigma}_{\boldsymbol{FY}}\boldsymbol{\Psi}(t) - \frac{\gamma^2}{2}\boldsymbol{D}_{\boldsymbol{\Gamma}}^{-1}.
\end{aligned}$$

Besides, we can check that $\boldsymbol{g}_0, f_0$ solve the corresponding ODEs for $\boldsymbol{\Gamma}$-neutral constraint.

As for matrix $\boldsymbol{\Lambda}(t)$, we have

$$\begin{aligned}
\boldsymbol{\Lambda}'(t) &= -\frac{1}{2}\boldsymbol{\Psi}(t)^\top \boldsymbol{\Sigma}\boldsymbol{\Psi}(t) + \frac{\gamma}{2}\boldsymbol{\Psi}(t)^\top \boldsymbol{\Sigma}_{\boldsymbol{FY}}\boldsymbol{\Sigma}_{\boldsymbol{F}}^{-1}\boldsymbol{\Gamma}\boldsymbol{D}_{\Gamma}^{-1} \\
&\quad + \frac{\gamma}{2}\boldsymbol{D}_{\Gamma}^{-1}\boldsymbol{\Gamma}^\top \boldsymbol{\Sigma}_{\boldsymbol{F}}^{-1}\boldsymbol{\Sigma}_{\boldsymbol{FY}}^\top \boldsymbol{\Psi}(t) - \frac{\gamma^2}{2}\boldsymbol{D}_{\Gamma}^{-1} \\
&\leq -\frac{1}{2}\boldsymbol{\Psi}(t)^\top \boldsymbol{\Sigma}_{\boldsymbol{FY}}^\top \boldsymbol{\Sigma}_{\Gamma}\boldsymbol{\Sigma}_{\boldsymbol{FY}}\boldsymbol{\Psi}(t) \\
&\quad + \frac{\gamma}{2}\boldsymbol{\Psi}(t)^\top \boldsymbol{\Sigma}_{\boldsymbol{FY}}^\top \boldsymbol{\Sigma}_{\boldsymbol{F}}^{-1}\boldsymbol{\Gamma}\boldsymbol{D}_{\Gamma}^{-1} + \frac{\gamma}{2}\boldsymbol{D}_{\Gamma}^{-1}\boldsymbol{\Gamma}^\top \boldsymbol{\Sigma}_{\boldsymbol{F}}^{-1}\boldsymbol{\Sigma}_{\boldsymbol{FY}}\boldsymbol{\Psi}(t) - \frac{\gamma^2}{2}\boldsymbol{D}_{\Gamma}^{-1} \\
&= -\frac{1}{2}\left(\boldsymbol{\Gamma}^\top \boldsymbol{\Sigma}_{\boldsymbol{F}}^{-1}\boldsymbol{\Sigma}_{\boldsymbol{FY}}\boldsymbol{\Psi}(t) + \gamma\right)^\top \boldsymbol{D}_{\Gamma}^{-1}\left(\boldsymbol{\Gamma}^\top \boldsymbol{\Sigma}_{\boldsymbol{F}}^{-1}\boldsymbol{\Sigma}_{\boldsymbol{FY}}\boldsymbol{\Psi}(t) + \gamma\right),
\end{aligned}$$

where the first inequality comes from equation (6.26) that

$$\begin{aligned}
\boldsymbol{\Sigma} &= \boldsymbol{\Sigma}_{\boldsymbol{Y}} - \boldsymbol{\Sigma}_{\boldsymbol{FY}}^\top(\boldsymbol{\Sigma}_{\boldsymbol{F}}^{-1} - \boldsymbol{\Sigma}_{\Gamma})\boldsymbol{\Sigma}_{\boldsymbol{FY}} \\
&\geq \boldsymbol{\Sigma}_{\boldsymbol{FY}}^\top \boldsymbol{\Sigma}_{\Gamma}\boldsymbol{\Sigma}_{\boldsymbol{FY}}.
\end{aligned}$$

Hence, we have shown that $\boldsymbol{\Lambda}'(t)$ is negative semidefinite, so $\boldsymbol{\Lambda}(t)$ is positive semidefinite as $\boldsymbol{\Lambda}(T) = \mathbf{0}$.

Proof of Proposition 6.11

The ODEs for the auxiliary functions

$$\begin{aligned}
\widetilde{\boldsymbol{H}}(t) &= \boldsymbol{H}^{no}(t) - \boldsymbol{H}(t), \\
\widetilde{\boldsymbol{g}}(t) &= \boldsymbol{g}^{no}(t) - \boldsymbol{g}(t), \\
\widetilde{f}(t) &= f^{no}(t) - f(t)
\end{aligned}$$

can be described in multiple ways. We opt to express them without involving the functions $\boldsymbol{H}^{no}(t), \boldsymbol{g}(t)$, or $f^{no}(t)$ as follows:

$$\begin{aligned}
\widetilde{\boldsymbol{H}}'(t) &= \left(\boldsymbol{K}(t) + \boldsymbol{K}(t)\boldsymbol{\Sigma}_{\boldsymbol{F}}^{-1}\boldsymbol{\Sigma}_{\boldsymbol{FY}} + \boldsymbol{H}(t)\boldsymbol{\Sigma}^{no}\right)\widetilde{\boldsymbol{H}}(t) \\
&\quad + \widetilde{\boldsymbol{H}}(t)\left(\boldsymbol{K}(t) + \boldsymbol{K}(t)\boldsymbol{\Sigma}_{\boldsymbol{F}}^{-1}\boldsymbol{\Sigma}_{\boldsymbol{FY}} + \boldsymbol{H}(t)\boldsymbol{\Sigma}^{no}\right)^\top \\
&\quad + \widetilde{\boldsymbol{H}}(t)\boldsymbol{\Sigma}^{no}\widetilde{\boldsymbol{H}}(t) - (\boldsymbol{H}(t)\boldsymbol{\Sigma}_{\boldsymbol{FY}}^\top - \boldsymbol{K}(t))\boldsymbol{\Sigma}_{\Gamma}(\boldsymbol{\Sigma}_{\boldsymbol{FY}}\boldsymbol{H}(t) - \boldsymbol{K}(t)),
\end{aligned} \tag{6.38}$$

$$
\begin{aligned}
\widetilde{\boldsymbol{g}}'(t) &= \boldsymbol{K}(t)\widetilde{\boldsymbol{g}}(t) + \widetilde{\boldsymbol{H}}(t)\boldsymbol{m} + \widetilde{\boldsymbol{H}}(t)\boldsymbol{\Sigma}^{no}\boldsymbol{g} + \widetilde{\boldsymbol{H}}\boldsymbol{\Sigma}^{no}\widetilde{\boldsymbol{g}} + \boldsymbol{H}\boldsymbol{\Sigma}^{no}\widetilde{\boldsymbol{g}} \\
&\quad + \boldsymbol{K}(t)\boldsymbol{\Sigma}_{\boldsymbol{F}}^{-1}\boldsymbol{\Sigma}_{\boldsymbol{FY}}\widetilde{\boldsymbol{g}} - \widetilde{\boldsymbol{H}}\boldsymbol{\Sigma}_{\boldsymbol{FY}}^{\top}\boldsymbol{\Sigma}_{\boldsymbol{F}}^{-1}\boldsymbol{\mu}_{\boldsymbol{F}} \\
&\quad - (\boldsymbol{H}(t)\boldsymbol{\Sigma}_{\boldsymbol{FY}}^{\top} - \boldsymbol{K}(t))\boldsymbol{\Sigma}_{\boldsymbol{\Gamma}}(\boldsymbol{\Sigma}_{\boldsymbol{FY}}\boldsymbol{g} + \boldsymbol{\mu}_{\boldsymbol{F}} - \gamma\boldsymbol{\Sigma}_{\boldsymbol{F}}\boldsymbol{\pi}), \\
\widetilde{f}'(t) &= -\boldsymbol{m}^{\top}\widetilde{\boldsymbol{g}}(t) + \frac{1}{2}tr(\boldsymbol{\Sigma}_{\boldsymbol{Y}}\widetilde{\boldsymbol{H}}(t)) - \frac{1}{2}\widetilde{\boldsymbol{g}}(t)^{\top}\boldsymbol{\Sigma}^{no}\widetilde{\boldsymbol{g}}(t) \\
&\quad - \boldsymbol{g}^{\top}\boldsymbol{\Sigma}^{no}\widetilde{\boldsymbol{g}} + \boldsymbol{\mu}_{\boldsymbol{F}}^{\top}\boldsymbol{\Sigma}_{\boldsymbol{F}}^{-1}\boldsymbol{\Sigma}_{\boldsymbol{FY}}\widetilde{\boldsymbol{g}} \\
&\quad + \frac{1}{2}\left(\boldsymbol{\theta} + \boldsymbol{\Sigma}_{\boldsymbol{FY}}\boldsymbol{g}(t) + \gamma\boldsymbol{\Sigma}_{\boldsymbol{F}}\boldsymbol{\pi}\right)^{\top}\boldsymbol{\Sigma}_{\boldsymbol{\Gamma}}\left(\boldsymbol{\theta} + \boldsymbol{\Sigma}_{\boldsymbol{FY}}\boldsymbol{g}(t) + \gamma\boldsymbol{\Sigma}_{\boldsymbol{F}}\boldsymbol{\pi}\right), \\
\widetilde{\boldsymbol{H}}'(T) &= \boldsymbol{0}_{N\times N}, \\
\widetilde{\boldsymbol{g}}'(T) &= \boldsymbol{0}_{N\times 1}, \\
\widetilde{f}'(T) &= 0.
\end{aligned}
$$

Now, we define the matrix function

$$
\boldsymbol{M}(t) = \begin{pmatrix} \widetilde{\boldsymbol{H}}(t), & -\widetilde{\boldsymbol{g}}(t) \\ -\widetilde{\boldsymbol{g}}(t)^{\top}, & -2\widetilde{f}(t) \end{pmatrix}.
$$

Our goal is to show that this matrix function satisfies a proper Riccati differential equation by direct calculation. Precisely, we have

$$
\begin{aligned}
\boldsymbol{M}'(t) &= \boldsymbol{M}(t)\begin{pmatrix} \boldsymbol{\Sigma}^{no}, & \boldsymbol{0} \\ \boldsymbol{0}, & 0 \end{pmatrix}\boldsymbol{M}(t) \\
&\quad + \begin{pmatrix} \boldsymbol{K}(t) + \boldsymbol{H}\boldsymbol{\Sigma}^{no} + \boldsymbol{K}(t)\boldsymbol{\Sigma}_{\boldsymbol{F}}^{-1}\boldsymbol{\Sigma}_{\boldsymbol{FY}}, & \boldsymbol{0} \\ -\boldsymbol{m}^{\top} - \boldsymbol{g}^{\top}\boldsymbol{\Sigma}^{no} + \boldsymbol{\mu}_{\boldsymbol{F}}^{\top}\boldsymbol{\Sigma}_{\boldsymbol{F}}^{-1}\boldsymbol{\Sigma}_{\boldsymbol{FY}}, & 0 \end{pmatrix}\boldsymbol{M}(t) \\
&\quad + \boldsymbol{M}(t)\begin{pmatrix} \boldsymbol{K}(t) + \boldsymbol{H}\boldsymbol{\Sigma}^{no} + \boldsymbol{K}(t)\boldsymbol{\Sigma}_{\boldsymbol{F}}^{-1}\boldsymbol{\Sigma}_{\boldsymbol{FY}}, & \boldsymbol{0} \\ -\boldsymbol{m}^{\top} - \boldsymbol{g}^{\top}\boldsymbol{\Sigma}^{no} + \boldsymbol{\mu}_{\boldsymbol{F}}^{\top}\boldsymbol{\Sigma}_{\boldsymbol{F}}^{-1}\boldsymbol{\Sigma}_{\boldsymbol{FY}}, & 0 \end{pmatrix}^{\top} + \begin{pmatrix} Q_1, & Q_2 \\ Q_2^{\top}, & Q_4 \end{pmatrix},
\end{aligned}
\tag{6.39}
$$

where

$$
\begin{aligned}
Q_1 &= -(\boldsymbol{\Sigma}_{\boldsymbol{FY}}\boldsymbol{H}(t) - \boldsymbol{K}(t))^{\top}\boldsymbol{\Sigma}_{\boldsymbol{\Gamma}}(\boldsymbol{\Sigma}_{\boldsymbol{FY}}\boldsymbol{H}(t) - \boldsymbol{K}(t)), \\
Q_2 &= (\boldsymbol{H}(t)\boldsymbol{\Sigma}_{\boldsymbol{FY}}^{\top} - \boldsymbol{K}(t))\boldsymbol{\Sigma}_{\boldsymbol{\Gamma}}(\boldsymbol{\Sigma}_{\boldsymbol{FY}}\boldsymbol{g} + \boldsymbol{\mu}_{\boldsymbol{F}} - \gamma\boldsymbol{\Sigma}_{\boldsymbol{F}}\boldsymbol{\pi}), \\
Q_4 &= -tr(\boldsymbol{\Sigma}_{\boldsymbol{Y}}\widetilde{\boldsymbol{H}}) - \left(\boldsymbol{\theta} + \boldsymbol{\Sigma}_{\boldsymbol{FY}}\boldsymbol{g}(t) - \gamma\boldsymbol{\Sigma}_{\boldsymbol{F}}\boldsymbol{\pi}\right)^{\top}\boldsymbol{\Sigma}_{\boldsymbol{\Gamma}}\left(\boldsymbol{\theta} + \boldsymbol{\Sigma}_{\boldsymbol{FY}}\boldsymbol{g}(t) + \gamma\boldsymbol{\Sigma}_{\boldsymbol{F}}\boldsymbol{\pi}\right).
\end{aligned}
$$

Moreover, the differential question (6.38) implies that the matrix $\widetilde{\boldsymbol{H}}$ is positive semidefinite. Therefore, there exists a matrix B such that

$$\widetilde{\boldsymbol{H}} = B^\top B.$$

Then, we can write

$$\begin{aligned} -tr(\boldsymbol{\Sigma_Y} \widetilde{\boldsymbol{H}}) &= -tr(B\boldsymbol{\Sigma_Y} B^\top) \\ &\leq 0. \end{aligned}$$

Using this inequality for the last term in (6.39), we have

$$\begin{aligned} \begin{pmatrix} Q_1, & Q_2 \\ Q_2^\top, & Q_4 \end{pmatrix} = \begin{pmatrix} \mathbf{0}, & \mathbf{0} \\ \mathbf{0}, & -tr(\boldsymbol{\Sigma_Y}\widetilde{\boldsymbol{H}}) \end{pmatrix} - \begin{pmatrix} \boldsymbol{K}(t) - \boldsymbol{H}(t)\boldsymbol{\Sigma}_{\boldsymbol{FY}}^\top \\ \boldsymbol{\theta}^\top + \boldsymbol{g}(t)^\top \boldsymbol{\Sigma}_{\boldsymbol{FY}}^\top - \gamma\boldsymbol{\pi}^\top \boldsymbol{\Sigma_F} \end{pmatrix} \boldsymbol{\Sigma_\Gamma} \\ \times \begin{pmatrix} \boldsymbol{K}(t) - \boldsymbol{H}(t)\boldsymbol{\Sigma}_{\boldsymbol{FY}}^\top \\ \boldsymbol{\theta}^\top + \boldsymbol{g}(t)^\top \boldsymbol{\Sigma}_{\boldsymbol{FY}}^\top - \gamma\boldsymbol{\pi}^\top \boldsymbol{\Sigma_F} \end{pmatrix}^\top, \end{aligned}$$

and it is negative semidefinite. In addition, $\boldsymbol{\Sigma}^{no}$ is positive semidefinite. Therefore, $\boldsymbol{M}(t)$ is the corresponding unique positive semidefinite solution for the Riccati equation (6.39).

Proof of Theorem 6.13

Let u be the solution given in Theorem 6.3. To prove this theorem, we first establish the following two assertions:

(a) With any admissible strategy $\boldsymbol{\pi} \in \mathcal{A}$, satisfying linear constraints $\boldsymbol{\Gamma}^\top \boldsymbol{\pi} = \boldsymbol{c}$, we have

$$u(t, \boldsymbol{y}, w) \geq \mathbb{E}^{\mathbb{P}}_{t,\boldsymbol{y},w}[U(W_T^{\boldsymbol{\pi}})],$$

for all $(t, \boldsymbol{y}, w) \in [0, T] \times \mathbb{R}^N \times D$, where $\mathbb{E}^{\mathbb{P}}_{t,\boldsymbol{y},w}[\cdot]$ denotes the conditional expectation $\mathbb{E}^{\mathbb{P}}[\cdot | W_t^{\boldsymbol{\pi}} = w, \boldsymbol{Y}_t = \boldsymbol{y}]$ and $W_T^{\boldsymbol{\pi}}$ is the terminal wealth.

(b) There exists an admissible strategy $\boldsymbol{\pi}^* \in \mathcal{A}$, satisfying $\boldsymbol{\Gamma}^\top \boldsymbol{\pi}^* = \boldsymbol{c}$ such that

$$u(t, \boldsymbol{y}, w) = \mathbb{E}^{\mathbb{P}}_{t,\boldsymbol{y},w}[U(W_T^{\boldsymbol{\pi}^*})],$$

for $(t, \boldsymbol{y}, w) \in [0, T] \times \mathbb{R}^N \times D$.

Provided the two statements above are correct, then we have two opposite inequalities. Specifically, statement (a) implies that $u \geq V$, and statement (b) implies that $u \leq V$. Therefore, we obtain the equality $u = V$ as desired.

Proof. We now proceed to prove statements (a) and (b) above.

(a) Given $\pi \in \mathcal{A}$, satisfying the constraint $\boldsymbol{\Gamma}^\top \pi = \boldsymbol{c}$, we apply Ito's formula to $u(t, \boldsymbol{y}, w)$, to get

$$
\begin{aligned}
&du(t, \boldsymbol{Y}_t, W_t^{\boldsymbol{\pi}}) \\
&= \left\{ u_t + (\mathbf{m} - \boldsymbol{K}(t)\boldsymbol{Y}_t)\nabla_{\boldsymbol{y}} u + \frac{1}{2} tr(\boldsymbol{\Sigma_Y} \nabla_{\mathbf{y}}^2 u) \right. \\
&\quad \left. + \boldsymbol{\pi}^\top (\boldsymbol{\mu}_{\mathbf{F}} + \boldsymbol{K}(t)\boldsymbol{Y}_t) u_w + \boldsymbol{\pi}^\top \boldsymbol{\Sigma_{FY}} \nabla_{\boldsymbol{y}}(u_w) + \frac{1}{2}\boldsymbol{\pi}^\top \boldsymbol{\Sigma_F} \boldsymbol{\pi} \partial_{ww} u \right\} dt \\
&\quad + (\nabla_{\boldsymbol{y}} u)^\top \big(\widetilde{\boldsymbol{\Sigma}}_{\boldsymbol{YS}} d\boldsymbol{Z}_{t,1} + \widetilde{\boldsymbol{\Sigma}}_{\boldsymbol{Y}} d\boldsymbol{Z}_{t,2}\big) + u_w \boldsymbol{\pi}^\top \big(\widetilde{\boldsymbol{\Sigma}}_{\mathbf{FS}} d\boldsymbol{Z}_{t,1} + \widetilde{\boldsymbol{\Sigma}}_{\mathbf{F}} d\boldsymbol{Z}_{t,2}\big).
\end{aligned}
$$

According to the HJB equation (6.23) for $u(t, \boldsymbol{y}, w)$, we have

$$
\begin{aligned}
du(t, \boldsymbol{Y}_t, W_t^{\boldsymbol{\pi}}) &\leq (\nabla_{\boldsymbol{y}} u)^\top \big(\widetilde{\boldsymbol{\Sigma}}_{\boldsymbol{YS}} d\boldsymbol{Z}_{t,1} + \widetilde{\boldsymbol{\Sigma}}_{\boldsymbol{Y}} d\boldsymbol{Z}_{t,2}\big) \\
&\quad + u_w \boldsymbol{\pi}^\top \big(\widetilde{\boldsymbol{\Sigma}}_{\mathbf{FS}} d\boldsymbol{Z}_{t,1} + \widetilde{\boldsymbol{\Sigma}}_{\mathbf{F}} d\boldsymbol{Z}_{t,2}\big) \\
&= u(t, \boldsymbol{Y}_t, W_t^{\boldsymbol{\pi}}) \big((\boldsymbol{g}(t) - \boldsymbol{H}(t)\boldsymbol{Y}_t)^\top \big(\widetilde{\boldsymbol{\Sigma}}_{\boldsymbol{YS}} d\boldsymbol{Z}_{t,1} + \widetilde{\boldsymbol{\Sigma}}_{\boldsymbol{Y}} d\boldsymbol{Z}_{t,2}\big) \\
&\quad - \gamma \boldsymbol{\pi}^\top \big(\widetilde{\boldsymbol{\Sigma}}_{\mathbf{FS}} d\boldsymbol{Z}_{t,1} + \widetilde{\boldsymbol{\Sigma}}_{\mathbf{F}} d\boldsymbol{Z}_{t,2}\big)\big) \\
&= u(t, \boldsymbol{Y}_t, W_t^{\boldsymbol{\pi}}) dA_t^{\boldsymbol{\pi}}.
\end{aligned}
$$

Therefore, we obtain

$$
U(W_T^{\boldsymbol{\pi}}) \leq u(t, \boldsymbol{Y}_t, W_t^{\boldsymbol{\pi}}) \mathcal{E}(A_\cdot^{\boldsymbol{\pi}} - A_t^{\boldsymbol{\pi}})_T.
$$

Taking the conditional expectation for both sides completes the proof of assertion (a) above. Moreover, the equality holds when taking $\boldsymbol{\pi} = \boldsymbol{\pi}^*$.

(b) It suffices to show that $\boldsymbol{\pi}^*$ is admissible. We combine the integral form of $\boldsymbol{Y}_t$ according to Remark 3.5 in Angoshtari and Leung (2020) and integration by parts technique to check that $\boldsymbol{Y}_t$ satisfies Benes condition below, thus, the corresponding integrand of $A^{\boldsymbol{\pi}^*}$ also satisfies

Benes condition. That is, there exists some constant K such that

$$\|(-\boldsymbol{Y}_t^\top \boldsymbol{H}(t) + \boldsymbol{g}(t)^\top)\tilde{\boldsymbol{\Sigma}}_{\boldsymbol{YS}}\|_{L^1} \leq K(1 + \max_{0\leq s\leq t} \|(\boldsymbol{Z}_{s,1}, \boldsymbol{Z}_{s,2})\|_{L^1}),$$

$$\|(-\boldsymbol{Y}_t^\top \boldsymbol{H}(t) + \boldsymbol{g}(t)^\top)\tilde{\boldsymbol{\Sigma}}_{\boldsymbol{Y}}\|_{L^1} \leq K(1 + \max_{0\leq s\leq t} \|(\boldsymbol{Z}_{s,1}, \boldsymbol{Z}_{s,2})\|_{L^1}),$$

$$\|\boldsymbol{\pi}_t^{*\top} \tilde{\boldsymbol{\Sigma}}_{\boldsymbol{FS}}\|_{L^1} \leq K(1 + \max_{0\leq s\leq t} \|(\boldsymbol{Z}_{s,1}, \boldsymbol{Z}_{s,2})\|_{L^1}),$$

$$\|\boldsymbol{\pi}_t^{*\top} \tilde{\boldsymbol{\Sigma}}_{\boldsymbol{F}}\|_{L^1} \leq K(1 + \max_{0\leq s\leq t} \|(\boldsymbol{Z}_{s,1}, \boldsymbol{Z}_{s,2})\|_{L^1}),$$

where $\boldsymbol{\pi}^*$ is given by (6.31). See Beneš (1971) or Karatzas and Shreve (1991), which verifies the admissibility conditions. □

Bibliography

Adjemian, M. K., Garcia, P., Irwin, S. and Smith, A. (2013). Non-convergence in domestic commodity futures markets: Causes, consequences, and remedies, *US Department of Agriculture, Economic Research Service* **115**, p. 155381.

Almansour, A. (2016). Convenience yield in commodity price modeling: A regime switching approach, *Energy Economics* **53**, pp. 238–247.

Angoshtari, B. (2016). On the market-neutrality of optimal pairs-trading strategies, *ArXiv e-prints*.

Angoshtari, B. and Leung, T. (2019a). Optimal dynamic basis trading, *Annals of Finance* **15**, 3, pp. 307–335.

Angoshtari, B. and Leung, T. (2019b). Optimal dynamic basis trading, *Annals of Finance* **15**, 3, pp. 307–335.

Angoshtari, B. and Leung, T. (2020). Optimal trading of a basket of futures contracts, *Annals of Finance* **16**, 2, pp. 253–280.

Aragon, G. O., Mehra, R. and Wahal, S. (2020). Do properly anticipated prices fluctuate randomly? evidence from VIX futures markets, *The Journal of Portfolio Management* **46**, 7, pp. 144–159.

Barndorff-Nielsen, O. E., Benth, F. E. and Veraart, A. E. D. (2015). *Cross-Commodity Modelling by Multivariate Ambit Fields* (Springer New York, New York, NY), pp. 109–148.

Baur, D. G. and Smales, L. A. (2022). Trading behavior in bitcoin futures: Following the "smart money", *Journal of Futures Markets* **42**, 7, pp. 1304–1323.

Beneš, V. E. (1971). Existence of optimal stochastic control laws, *SIAM Journal on Control* **9**, 3, pp. 446–472, doi:10.1137/0309034.

Benth, F. E. and Karlsen, K. H. (2005). A note on Merton's portfolio selection problem for the Schwartz mean-reversion model, *Stochastic Analysis and Applications* **23**, 4, pp. 687–704.

Bhar, R. and Lee, D. (2011). Time-varying market price of risk in the crude oil futures market, *Journal of Futures Markets* **31**, 8, pp. 779–807.

Bodie, Z. and Rosansky, V. I. (1980). Risk and return in commodity futures, *Financial Analysts Journal* **36**, 3, pp. 27–39.

Brennan, M. J. and Schwartz, E. S. (1988). Optimal arbitrage strategies under basis variability, in M. Sarnat (ed.), *Essays in Financial Economics* (North Holland, New York), pp. 167–180.

Brennan, M. J. and Schwartz, E. S. (1990). Arbitrage in stock index futures, *Journal of Business* **63**, 1, pp. S7–S31.

Brody, D. C., Hughston, L. P. and Macrina, A. (2008). Information-based asset pricing, *International Journal of Theoretical and Applied Finance* **11**, 1, pp. 107–142.

Buetow, G. W. and Henderson, B. J. (2016). The VIX futures basis: Determinants and implications, *The Journal of Portfolio Management* **42**, 2, pp. 119–130.

Buffington, J. and Elliott, R. J. (2002). American options with regime switching, *International Journal of Theoretical and Applied Finance* **5**, pp. 497–514.

Capponi, A. and Figueroa-López (2012). Dynamic portfolio optimization with a defaultable security and regime-switching, *Mathematical Finance* **24**, pp. 207–249.

Carmona, R. and Ludkovski, M. (2004). Spot convenience yield models for energy markets, *Contemporary Mathematics* **351**, pp. 65–80.

Cartea, Á. and Jaimungal, S. (2016). Algorithmic trading of co-integrated assets, *International Journal of Theoretical and Applied Finance* **19**, 6, p. 1650038.

Cartea, Á., Jaimungal, S. and Kinzebulatov, D. (2016). Algorithmic trading with learning, *International Journal of Theoretical and Applied Finance* **19**, 4, p. 1650028.

Çanakoğlu, E. and Özekici (2012). HARA frontiers of optimal portfoilos in stochastic markets, *European Journal of Operational Research* **221**, pp. 129–137.

Chen, K., Chiu, M. and Wong, H. (2019). Time-consistent mean-variance pairs-trading under regime-switching cointegration, *SIAM Journal on Financial Mathematics* **10**, 2, pp. 632–665, doi:10.1137/18M1209611.

Chen, S., Li, Q., Wang, Q. and Zhang, Y. Y. (2023). Multivariate models of commodity futures markets: A dynamic copula approach, *Empirical Economics* **64**, pp. 3037–3057, https://doi.org/10.1007/s00181-023-02373-2.

Chen, X., Leung, T. and Zhou, Y. (2022). Constrained dynamic futures portfolios with stochastic basis, *Annals of Finance* **18**, pp. 1–33.

Cortazar, G., Lopez, M. and Naranjo, L. (2017). A multifactor stochastic volatility model of commodity prices, *Energy Economics* **67**, pp. 182–201.

Cortazar, G., Milla, C. and Severino, F. (2008). A multicommodity model of futures prices: Using futures prices of one commodity to estimate the stochastic process of another, *Journal of Futures Markets* **28**, 6, pp. 537–560.

Cortazar, G., Millard, C., Ortega, H. and Schwartz, E. S. (2019). Commodity price forecasts, futures prices, and pricing models, *Management Science* **65**, 9, pp. 4141–4155.

Cortazar, G. and Naranjo, L. (2006). An N-factor Gaussian model of oil futures prices, *Journal of Futures Markets* **26**, 3, pp. 243–268.

Cummings, J. R. and Frino, A. (2011). Index arbitrage and the pricing relationship between Australian stock index futures and their underlying shares, *Accounting & Finance* **51**, 3, pp. 661–683.

Dai, M., Zhong, Y. and Kwok, Y. K. (2011). Optimal arbitrage strategies on stock index futures under position limits, *Journal of Futures Markets* **31**, 4, pp. 394–406.

Demiralay, S., Bayraci, S. and Gaye Gencer, H. (2019). Time-varying diversification benefits of commodity futures, *Empirical Economics* **56**, 6, pp. 1823–1853.

Deng, J., Pan, H., Zhang, S. and Zou, B. (2020). Minimum-variance hedging of bitcoin inverse futures, *Applied Economics* **52**, 58, pp. 6320–6337.

Elaut, G., Erdös, P. and Sjödin, J. (2016). An analysis of the risk-return characteristics of serially correlated managed futures, *Journal of Futures Markets* **36**, 10, pp. 992–1013.

Elliott, R. J., Chan, L. and Siu, T. K. (2005). Option pricing and esscher transform under regime switching, *Annals of Finance* **1**, pp. 423–432.

Ewald, C.-O., Zhang, A. and Zong, Z. (2019). On the calibration of the Schwartz two-factor model to WTI crude oil options and the extended Kalman filter, *Annals of Operations Research* **282**, 1, pp. 119–130.

Fleming, W. and Soner, H. (1993). *Controlled Markov Processes and Viscosity Solutions* (Springer, New York).

Garcia, P., Irwin, S. H. and Smith, A. (2015). Futures market failure? *American Journal of Agricultural Economics* **97**, 1, pp. 40–64.

Gómez-Valle, L. and Martínez-Rodríguez, J. (2013). Advances in pricing commodity futures: Multifactor models, *Mathematical and Computer Modelling* **57**, 7, pp. 1722–1731.

Gregoriou, G. N., Hubner, G. and Kooli, M. (2010). Performance and persistence of commodity trading advisors: Further evidence, *Journal of Futures Markets* **30**, 8, pp. 725–752.

Guijarro-Ordonez, J. (2019). Stochastic control in high-dimensional statistical arbitrage under an Ornstein-Uhlenbeck process, *Applied Mathematical Finance* **26**, 4, pp. 328–358.

Guijarro-Ordonez, J., Pelger, M. and Zanotti, G. (2022). Deep learning statistical arbitrage, Working paper, arXiv:2106.04028.

Guo, K. and Leung, T. (2017). Understanding the non-convergence of agricultural futures via stochastic storage costs and timing options, *Journal of Commodity Markets* **6**, pp. 32–49.

Guo, K., Leung, T. and Ward, B. (2019). How to mine gold without digging, *International Journal of Financial Engineering* **6**, 1, p. 1950009.

Hamilton, J. (1989). A new approach to the economic analysis of nonstationary time series and the business cycle, *Econometrica* **57**, pp. 357–384.

Hurst, B., Ooi, Y.-H. and Pedersen, L. H. (2013). Demystifying managed futures, *Journal of Investment Management* **11**, 3, pp. 42–58.

Irwin, S. H., Garcia, P., Good, D. L. and Kunda, E. L. (2011). Spreads and non-convergence in Chicago board of trade corn, soybean, and wheat futures: Are index funds to blame? *Applied Economic Perspectives and Policy* **33**, 1, pp. 116–142.

Jackson, K. R., Jaimungal, S. and Surkov, V. (2008). Fourier space time-stepping for option pricing with Lévy models, *Journal of Computational Finance* **12**, 2, pp. 1–29.

Karatzas, I. and Shreve, S. (1991). *Brownian Motion and Stochastic Calculus* (Springer-Verlag, New York).

Kuroda, K. and Nagai, H. (2002). Risk-sensitive portfolio optimization on infinite time horizon, *Stochastics and Stochastics Reports* **73**, pp. 309–331.

Kurov, A. (2008). Investor sentiment, trading behavior and informational efficiency in index futures markets, *Financial Review* **43**, 1, pp. 107–127.

Lee, K., Leung, T. and Ning, B. (2023). A diversification framework for multiple pairs trading strategies, *Risks* **11**, 5, p. 93.

Leung, T. (2010). A Markov-modulated stochastic control problem with optimal stopping with application to finance, in *Proceedings of the 49th IEEE Conference on Decision and Control*, pp. 559–566.

Leung, T., Li, J. and Li, X. (2018). Optimal timing to trade along a randomized Brownian bridge, *International Journal of Financial Studies* **6**, 3, p. 75.

Leung, T., Li, J., Li, X. and Wang, Z. (2016). Speculative futures trading under mean reversion, *Asia-Pacific Financial Markets* **23**, 4, pp. 281–304.

Leung, T. and Li, X. (2016). *Optimal Mean Reversion Trading: Mathematical Analysis and Practical Applications*, Modern Trends in Financial Engineering (World Scientific Publishing Company, Singapore).

Leung, T. and Park, H. (2017). Long-term growth rate of expected utility for leveraged ETFs: Martingale extraction approach, *International Journal of Theoretical and Applied Finance* **20**, 6, p. 1750037.

Leung, T., Park, H. and Yeo, H. (2024). Robust long-term growth rate of expected utility for leveraged ETFs. *Mathematics and Financial Economics.* https://doi.org/10.1007/s11579-024-00371-1.

Leung, T. and Wan, H. (2015). ESO valuation with job termination risk and jumps in stock price, *SIAM Journal on Financial Mathematics* **6**, 1, pp. 487–516.

Leung, T. and Ward, B. (2015). The golden target: Analyzing the tracking performance of leveraged gold ETFs, *Studies in Economics and Finance* **32**, 3, pp. 278–297.

Leung, T. and Ward, B. (2018). Dynamic index tracking and risk exposure control using derivatives, *Applied Mathematical Finance* **25**, 2, pp. 180–212.

Leung, T. and Ward, B. (2022). *Tracking VIX with VIX Futures: Portfolio Construction and Performance* (World Scientific, Singapore), pp. 557–596.

Leung, T. and Yan, R. (2018). Optimal dynamic pairs trading of futures under a two-factor mean-reverting model, *International Journal of Financial Engineering* **5**, 3, p. 1850027.

Leung, T. and Yan, R. (2019). A stochastic control approach to managed futures portfolios, *International Journal of Financial Engineering* **6**, 1, p. 1950005.

Leung, T., Zhang, J. and Aravkin, A. (2020). Sparse mean-reverting portfolios via penalized likelihood optimization, *Automatica* **111**, p. 108651.

Leung, T. and Zhou, Y. (2019). Dynamic optimal futures portfolio in a regime-switching market framework, *International Journal of Financial Engineering* **6**, 4, p. 1950034.

Li, T. N. and Papanicolaou, A. (2019). Dynamic optimal portfolios for multiple co-integrated assets, *Working paper.*

Liu, J. and Longstaff, F. A. (2004). Losing money on arbitrage: Optimal dynamic portfolio choice in markets with arbitrage opportunities, *The Review of Financial Studies* **17**, 3, pp. 611–641.

Liu, J. and Timmermann, A. (2013). Optimal convergence trade strategies, *Review of Financial Studies* **26**, 4, pp. 1048–1086.

Lubnau, T. and Todorova, N. (2015). Trading on mean-reversion in energy futures markets, *Energy Economics* **51**, pp. 312–319.

Meade, N. (2010). Oil prices - Brownian motion or mean reversion? A study using a one year ahead density forecast criterion, *Energy Economics* **32**, 6, pp. 1485–1498.

Mencia, J. and Sentana, E. (2013). Valuation of VIX derivatives, *Journal of Financial Economics* **108**, pp. 367–391.

Merton, R. (1971). Optimum consumption and portfolio rules in a continuous time model, *Journal of Economic Theory* **3**, pp. 373–413.

Miffre, J. (2016). Long-short commodity investing: A review of the literature, *Journal of Commodity Markets* **1**, 1, pp. 3–13.

Novikov, A. A. (1972). On an identity for stochastic integrals, *Theory of Probability & Its Applications* **17**, 4.

Oksendal, B. (2014). *Stochastic Differential Equations: An Introduction with Applications*, 6th edn. (Springer, Berlin, Germany).

Patton, A. J. (2008). Are "market neutral" hedge funds really market neutral? *The Review of Financial Studies* **22**, 7, pp. 2495–2530.

Richie, N., Daigler, R. T. and Gleason, K. C. (2008). The limits to stock index arbitrage: Examining S&P 500 futures and SPDRS, *Journal of Futures Markets* **28**, 12, pp. 1182–1205, doi:https://doi.org/10.1002/fut.20365.

Ross, K. (2008). *Stochastic Control in Continuous Time* (Lecture Notes on Continuous Time Stochastic Control).

Sass, J. and Haussmann, U. G. (2004). Optimizing the terminal wealth under partial information: The drift process as a continuous time Markov chain, *Finance and Stochastics* **8**, pp. 553–577.

Schwartz, E. (1997). The stochastic behavior of commodity prices: Implications for valuation and hedging, *The Journal of Finance* **52**, 3, pp. 923–973.

Schwartz, E. and Smith, J. E. (2000). Short-term variations and long-term dynamics in commodity prices, *Management Science* **46**, 7, pp. 893–911.

Simon, D. P. and Campasano, J. (2014). The VIX futures basis: Evidence and trading strategies, *The Journal of Derivatives* **21**, 3, pp. 54–69.

Smales, L. A. (2016). Trading behavior in S&P 500 index futures, *Review of Financial Economics* **28**, 1, pp. 46–55.

Sotomayor, L. R. and Cadenillas, A. (2009). Explicit solutions of consumption-investment problems in financial markets with regime switching, *Mathematical Finance* **19**, 2, pp. 251–279.

Surkov, V. (2009). *Option Pricing Using Fourier Space Time-Stepping Framework*, Ph.D. thesis, University of Toronto.

Trujillo-Barrera, A., Mallory, M. and Garcia, P. (2012). Volatility spillovers in U.S. crude oil, ethanol, and corn futures markets, *Journal of Agricultural and Resource Economics* **37**, 2, pp. 247–262.

Tsuji, C. (2018). Return transmission and asymmetric volatility spillovers between oil futures and oil equities: New DCC-MEGARCH analyses, *Economic Modelling* **74**, pp. 167–185.

Valle, C. and N. Meade, J. B. (2014). Market neutral portfolios, *Optimization Letters* **8**, 7, pp. 1961–1984.

Vega, C. A. M. (2018). Calibration of the exponential Ornstein–Uhlenbeck process when spot prices are visible through the maximum log-likelihood method: Example with gold prices, *Advances in Difference Equations* **269**, https://doi.org/10.1186/s13662-018-1718-4.

Waldow, F., Schnaubelt, M., Krauss, C. and Fischer, T. G. (2021). Machine learning in futures markets, *Journal of Risk and Financial Management* **14**, 3, doi:10.3390/jrfm14030119.

Wang, L., Ahmad, F., Luo, G.-l., Umar, M. and Kirikkaleli, D. (2022). Portfolio optimization of financial commodities with energy futures, *Annals of Operations Research* **313**, 1, pp. 401–439.

Zaremba, A., Bianchi, R. J. and Mikutowski, M. (2021). Long-run reversal in commodity returns: Insights from seven centuries of evidence, *Journal of Banking & Finance* **133**, p. 106238.

Zhao, Z. and Palomar, D. P. (2018). Mean-reverting portfolio with budget constraint, *IEEE Transactions on Signal Processing* **66**, 9, pp. 2342–2357.

Zhou, X. and Yin, G. (2003). Markowitz's mean-variance portfolio selection with regime switching: A continuous-time model, *SIAM Journal on Control and Optimization* **42**, 4, pp. 1466–1482.

Index

www.ingramcontent.com/pod-product-compliance
Lightning Source LLC
Chambersburg PA
CBHW050914171224
19114CB00004B/98

9 789811 282744